Lecture Notes in Mathematics

Edited by A. Dold and B. Eckmann

Subseries: Mathematisches Institut der Universität und Max-Planck-Institut für Mathematik, Bonn

Adviser: F. Hirzebruch

1062

Jürgen Jost

Harmonic Maps Between Surfaces

(with a Special Chapter on Conformal Mappings)

Springer-Verlag

Berlin Heidelberg New York Tokyo 1984

Author

Jürgen Jost
Mathematisches Institut der Universität
Wegelerstr. 10, 5300 Bonn, Federal Republic of Germany

AMS Subject Classification (1980): 58E20; 30C70, 32G15, 35J60

ISBN 3-540-13339-9 Springer-Verlag Berlin Heidelberg New York Tokyo
ISBN 0-387-13339-9 Springer-Verlag New York Heidelberg Berlin Tokyo

Printing and binding: Beltz Offsetdruck, Hemsbach/Bergstr.
2146/3140-543210

Dedicated to the memory of Dieter Kieven

PREFACE

The purpose of these Lecture Notes is twofold. On one hand, I want to give a fairly complete and self-contained account of the results on harmonic maps between surfaces. On the other hand, these notes should also serve as an introduction to the theory of harmonic maps in general; therefore, whenever appropriate, I point out which of the two-dimensional results pertain to higher dimensions and which do not, and I try to give some references and an idea of the respective proof. For a more complete account in this direction, however, the reader should consult the several excellent survey articles of Eells and Lemaire.

An essential aim of this book is to show the variety of methods and the interplay of different fields in the theory of harmonic maps, in particular the calculus of variations, partial differential equations, differential geometry, algebraic topology, and complex analysis. Thus, the concept of this book is strongly opposed to the view of a mere specialist. In particular, I think that a completely unified treatment of the topic is neither possible nor desirable.

Nevertheless, I believe that this treatment contains several simplifications and unifications compared to the presentations available in the existing literature. This book is not intended as a mere enumeration of unrelated results. On the contrary, the sequence of the chapters also reflects a logical order, and many different tools have to be constructed, until the results of the three final chapters can be proved. In particular, conformal mappings are used in a much more thorough way than in the existing literature. An outline of the contents now follows.

After giving an account of the history and presenting the definition of harmonic maps from several points of view in chapter 1, we start in chapter 2 with some geometric considerations. These concern convex discs on surfaces, and the result is roughly that if on a disc there are no conjugate points then there are also no cut points.

Moreover, we show the existence of local coordinates with curvature controlled Christoffel symbols, following Jost-Karcher [JK1].

Chapter 3 deals with conformal mappings. We first prove Theorem 9.3 in Morrey, "Multiple Integrals...", Springer, 1966, since Morrey's proof contains a mistake. The difficulty which leads to this error is overcome by minimizing energy in a restricted subclass of the Sobolev space H_2^1 which is suitably adapted to the problem, so that we can nevertheless conclude that the minimum is a conformal map with the desired properties (we shall encounter a similar idea in chapters 4 and 11). Furthermore, we shall prove that this map is a global diffeomorphism and as regular as one could expect (A-priori estimates, however, will only be obtained in later chapters).

In chapter 4, we first solve the Dirichlet problem for the case that the boundary values lie in some convex ball, a result due to Hildebrandt-Kaul-Widman [HKW 3]. Our proof consists of a combination of a rather general maximum principle and a lemma due to Courant and Lebesgue which is only valid in two dimensions. The proof also gives a-priori estimates for the modulus of continuity of the harmonic map. We then attack the general existence problem for harmonic maps between compact surfaces. Using the Courant-Lebesgue Lemma again, it is not hard to see that the limit of an energy minimizing can fall out of a homotopy class only if a sphere splits off. If the second homotopy group of the image vanishes, this cannot happen, however, and we thus obtain a new proof of the fundamental existence theorems of Lemaire [L1], [L2]. Furthermore, by a careful replacement argument, we can also solve the Dirichlet problem in two different homotopy classes for nonconstant boundary values, if the image is homeomorphic to a 2-sphere. In chapter 5, we deal with the question of uniqueness of harmon maps and prove the corresponding results of Hartmann [Ht] and Jäger-Kaul [JäK1] and then examine in some more detail the case of maps between closed surfaces.

In chapter 6, we prove $C^{1,\alpha}$-a-priori-estimates for the case where the domain is the unit disc in the plane. This latter assumption can then be removed with the help of the results of chapter 7 where we prove estimates for the functional determinant from below for univalent harmonic mappings between surfaces. These estimates apply in particular to conformal maps, and since a conformal map composed with a harmonic one is again harmonic, we can use the result of chapter 3 to pass from the unit disc to an arbitrary domain in chapter 9. The results of chapter 6 and 7 are based on Jost-Karcher [JK1] and employ several important ideas of E. Heinz.

In chapter 8, we prove the existence of harmonic diffeomorphisms as solutions of the Dirichlet problem, if the boundary values map the boundary of the domain homeomorphically onto a convex curve inside a convex disc. This result is taken from [J3] and uses in particular the results of chapter 7.

We can also use the a-priori-estimates to provide non-variational proofs of Thms. 4.1 and 8.1 in chapter 9, using Leray-Schauder degree theory.

We then apply Theorem 8.1 in chapter 10 to prove the existence of harmonic coordinates on arbitrary discs on a surface according to Jost-Karcher [JK1]. These coordinates possess best possible regularity properties and can be used to prove $C^{2,\alpha}$-a-priori estimates for harmonic maps between surfaces depending only on curvature bounds and injectivity radii, once the modulus of continuity is known.

Theorem 8.1 will again be applied in chapter 11 where we prove the existence of harmonic diffeomorphisms between closed surfaces, due to Jost-Schoen [JS]. We minimize energy in the class of diffeomorphisms and then apply a rather delicate replacement argument to show that the limit is a harmonic diffeomorphism.

The final chapter gives some applications of harmonic maps between surfaces. First, we give the analytic proof of Eells-Wood of a well known result of Kneser concerning mappings between closed surfaces, and then we give some applications of Earle-Eells and Tromba of harmonic maps to Teichmüller theory.

Furthermore, we discuss the Theorem of Ruh-Vilms stating that the Gauss map of a submanifold of Euclidean space with constant mean curvature is harmonic,as well as immersed surfaces in 3-space of constant Gauss curvature.

Among the omissions of the present book are results on the explicit construction and classification of harmonic maps between manifolds with canonical metrics. We refer the reader to [L 1], [EW 2], [EW 3], [EL 3] instead, since we do not feel that we can contribute anything new to the presentation of this area.

My work is indebted to several persons. To Hermann Karcher, I owe many insights into the geometric aspects of the field which he generously communicated to me. Furthermore, I benefitted much from collaboration or conversations with Jim Eells, Bob Gulliver, Luc Lemaire, Rick Schoen (in particular, Chapter 11 represents joint work with him), John Wood and Shing-Tung Yau. But most of all, I am indebted to Stefan Hildebrandt for his continous advice and encouragement over many years, and for supporting my research in every possible way through the means of the Sonderforschungsbereich 72 at the University of Bonn. Finally, I am grateful to Alfred Baldes for some useful comments on my manuscript and to Monika Zimmermann for typing it with great care and patience.

Table of contents

1. Introduction

1.1. A short history of variational principles

Among the first persons to realize the importance of variational problems and the physical significance of their solutions was G. W. Leibniz (1646 - 1716). In his work, however, mathematical and physical reasoning was closely interwoven with philosophical and theological arguments. One of the aims of his philosophy was to solve the problem of theodizee, i.e. to reconcile the evil in the world with God's goodness and almightiness (cf. [Lz]). Leibniz' answer was that God has chosen from the innumerable possible worlds the best possible, but that a perfect world is not possible. (This infinite multitude can only be conceived by an infinite understanding, which provided a proof of the existence of God for Leibniz.) This best possible world is distinguished by a pre - establishe harmony between itself, the kingdom of nature, on one hand and the heavenly kingdom of grace and freedom on the other hand. Through this the effective causes unite with the purposive causes. Thus bodies move due to their own internal laws in accordance with the thoughts and desires of the soul. In this way, the contradiction between the predetermination of the physical world following strict laws and the constantly experienced spontaneity and freedom of the individual is removed. The best possible world must here obey specific laws since an ordered world is better than a chaotic one. This proves therefore the necessity of the existence of natural laws. The contents of the natural laws, however, are not completely determined as is the case for geometric laws but are only determined in a moral sense, since they must satisfy the criteria of beauty and simplicity in the best of all possible worlds. This leads Leibniz even to variational principles. This is because if a physical process did not yield an extreme value, a maximum or minimum, for a particular energy or action integral, the world could be improved and would therefore not be the best possible one. Conversely, Leibniz also uses the beauty and simplicity of natural laws as evidence for his thesis of pre - established harmony. (The notion that we live in the best possible world was frequently rejected and even ridiculed by subsequent critics, in particular Voltaire, on account of the apparent flaws of this world, but Leibniz' point that a perfectly good world is not possible was beyond reach of these arguments.)

Leibniz, however, did not elaborate his argument concerning variational principles in his publications, but only in a private letter. Thus, it happened that a principle of least (and not only stationary) action was later rediscovered by Maupertuis (1698 - 1759), without knowing of Leibniz' idea. When S. König (1712 - 1757) then claimed priority for Leibniz on account of his letter that he was not able to show however to the Prussian Academy of Sciences (whose president was Maupertuis) this led to one of the most famous priority controversies in scientific history in which even Voltaire, Euler, and Frederick the Great became involved. It was also pointed out that Maupertuis' principle of least action should be replaced by a principle of stationary action since physical equilibria need only be stationary points but not necessarily minima of variational problems.

1.2. The concept of geodesics

One of the variational problems of most physical importance and mathematical interest was the problem of geodesics, i.e. to find the shortest (or at least locally shortest) connections between two points in a metric continuum, e.g. a Riemannian manifold. Applications of this concept range from general relativity, where geodesics describe the paths of moving bodies , to many innermathematical applications.
Geodesics are critical points of the length integral

$$\int_0^1 \left|\frac{\partial}{\partial t} c\right| dt$$

where $c:[0,1] \to N$ is the parametrization, as well as, if they are parametrized proportionally to arclength, of the energy integral

$$\int_0^1 \left|\frac{\partial}{\partial t} c\right|^2 dt.$$

Here, unfortunately, we find some ambiguity of terminology, since the mathematical term "energy" corresponds to the physical concept of "action", while in physics "energy" has a different meaning.
Because of the many applications of geodesics, it was rather natural to generalize this concept. While minimal surfaces are critical points of a twodimensional analogue of the length integral, namely the area integral, the generalization of the energy integral for maps between Riemannian manifolds led to the concept of harmonic maps. They are critical points of the corresponding integral where the squared norm of the gradient or energy density has to be defined in terms intrinsic to the geometry of the domain and target manifold and the map between them.

1.3. Definition of harmonic maps

Suppose that X and Y are Riemannian manifolds of dimensions n and N, resp., with metric tensors $(\gamma_{\alpha\beta})$ and (g_{ij}), resp., in some local coordinate charts $x = (x^1,\dots,x^n)$ and $f = (f^1,\dots,f^N)$ on X and Y, resp. Let $(\gamma^{\alpha\beta}) = (\gamma_{\alpha\beta})^{-1}$. If $f: X \to Y$ is a C^1-map, we can define the energy density

$$e(f) := \frac{1}{2}\gamma^{\alpha\beta}(x)\, g_{ij}(f)\, \frac{\partial f^i}{\partial x^\alpha}\frac{\partial f^j}{\partial x^\beta}$$

where we use the standard summation convention (greek minuscules occuring twice are summed from 1 to n, while latin ones are summed from 1 to N) and express everything in terms of local coordinates. Then the energy of f is simply

$$E(f) = \int_X e(f)\,dX$$

If f is of class C^2 and $E(f) < \infty$ and f is a critical point of E[1], then f is called harmonic and satisfies the corresponding Euler-Lagrange-equations. These are of the form

$$(1.3.1)\qquad \frac{1}{\sqrt{\gamma}}\frac{\partial}{\partial x^\alpha}\left(\sqrt{\gamma}\,\gamma^{\alpha\beta}\frac{\partial}{\partial x^\beta}f^i\right) + \gamma^{\alpha\beta}\Gamma^i_{jk}\frac{\partial}{\partial x^\alpha}f^j\frac{\partial}{\partial x^\beta}f^k = 0$$

in local coordinates, where $\gamma = \det(\gamma_{\alpha\beta})$ and the Γ^i_{jk} are the Christoffel symbols of the second kind on Y.
We thus obtain a nonlinear elliptic system of partial differential equations, where the principal part is the Laplace-Beltrami operator on X and is therefore in divergence form, while the nonlinearity is quadratic in the gradient of the solution.
We now want to look at the definition of harmonic maps from a more intrinsic point of view. The differential df of f, given in local coordinates by

$$df = \frac{\partial f^i}{\partial x^\alpha}\,dx^\alpha\,\frac{\partial}{\partial f^i}$$

can be considered as a section of the bundle $T^*X \otimes f^{-1}TY$. Then

$$e(f) = \frac{1}{2}\gamma^{\alpha\beta}\left\langle \frac{\partial f}{\partial x^\alpha}, \frac{\partial f}{\partial x^\beta}\right\rangle_{f^{-1}TY}$$

$$= \frac{1}{2}\langle df, df\rangle_{T^*X\otimes f^{-1}TY}$$

1) w.r.t. variations vanishing on ∂X, in case $\partial X \neq \emptyset$.

i.e. $e(f)$ is the trace of the pullback via f of the metric tensor of Y. In particular, $e(f)$ and hence also $E(f)$ are independant of the choice of local coordinates and thus intrinsically defined.
f is harmonic, if

$$(1.3.2) \qquad \tau(f) = 0 ,$$

where $\tau(f) = \operatorname{trace} \nabla df$, and ∇ here denotes the covariant derivative in the bundle $T^*X \otimes f^{-1} TY$.
Let us quickly show, why (1.3.1) and (1.3.2) are equivalent (cf. [EL 4]).

$$\nabla_{\frac{\partial}{\partial x^\beta}} (df) = \nabla_{\frac{\partial}{\partial x^\beta}} \left(\frac{\partial f^i}{\partial x^\alpha} dx^\alpha \frac{\partial}{\partial f^i}\right)$$
$$= \frac{\partial}{\partial x^\beta}\left(\frac{\partial f^i}{\partial x^\alpha}\right) dx^\alpha \frac{\partial}{\partial f^i} + \left(\nabla^{T^*X}_{\frac{\partial}{\partial x^\beta}} dx^\alpha\right) \frac{\partial f^i}{\partial x^\alpha} \frac{\partial}{\partial f^i} +$$
$$+ \left(\nabla^{f^{-1}TY}_{\frac{\partial}{\partial x^\beta}} \frac{\partial}{\partial f^i}\right) \frac{\partial f^i}{\partial x^\alpha} dx^\alpha =$$

$$= \frac{\partial^2 f^i}{\partial x^\alpha \partial x^\beta} dx^\alpha \frac{\partial}{\partial f^i} - {}^X\Gamma^\alpha_{\beta\gamma} dx^\gamma \frac{\partial f^i}{\partial x^\alpha} \frac{\partial}{\partial f^i} + {}^Y\Gamma^k_{ij} \frac{\partial}{\partial f^k} \frac{\partial f^j}{\partial x^\beta} \frac{\partial f^i}{\partial x^\alpha} dx^\alpha \text{ 1)}$$

and thus, since $\tau(f) = \operatorname{trace} \nabla df$,

$$\tau^k(f) = \gamma^{\alpha\beta} \frac{\partial^2 f^k}{\partial x^\alpha \partial x^\beta} - \gamma^{\alpha\beta}\, {}^X\Gamma^\gamma_{\alpha\beta} \frac{\partial f^k}{\partial x^\gamma} + \gamma^{\alpha\beta}\, {}^Y\Gamma^k_{ij} \frac{\partial f^i}{\partial x^\alpha} \frac{\partial f^j}{\partial x^\beta} ,$$

and we see that (1.3.1) and (1.3.2) are equivalent.
From the preceding calculation, we see that the Laplace - Beltrami operator is the contribution of the connection in T^*X, while the connection in $f^{-1} TY$ gives rise to the nonlinear term involving the Christoffel symbols of the image.
With the preceding notations, we can also calculate the Hessian of a harmonic map f for vector fields v, w along f (i.e. v and w are sections of $f^{-1} TY$). For this purpose, we consider a two - parameter variation f_{st} with

$$v = \frac{\partial f_{st}}{\partial s} \Big|_{s,t = 0} , \quad w = \frac{\partial f_{st}}{\partial t} \Big|_{s,t = 0}$$

1) Here, we distinguish the Christoffel symbpls of X and Y by the superscript X or Y , resp.

We then want to calculate

$$H_f(v,w) := \frac{\partial^2 E(f_{st})}{\partial s \partial t} \Big|\ s,t = 0$$

We have, writing f instead of f_{st}, and taking scalar products $\langle \cdot , \cdot \rangle$ in $T^*X \otimes f^{-1}TY$, if not otherwise indicated,

$$\frac{\partial}{\partial t}\frac{\partial}{\partial s}\frac{1}{2}\left\langle \frac{\partial f}{\partial x^\alpha}dx^\alpha, \frac{\partial f}{\partial x^\beta}dx^\beta \right\rangle =$$

$$= \frac{\partial}{\partial t}\left\langle \nabla_{\frac{\partial}{\partial s}} \frac{\partial f}{\partial x^\alpha}dx^\alpha , \frac{\partial f}{\partial x^\beta}dx^\beta \right\rangle =$$

$$= \frac{\partial}{\partial t}\left\langle \nabla^{f^{-1}TY}_{\frac{\partial}{\partial x^\alpha}} \left(\frac{\partial f}{\partial s}\right) dx^\alpha , \frac{\partial f}{\partial x^\beta}dx^\beta \right\rangle =$$

$$= \left\langle \nabla_{\frac{\partial}{\partial t}} \nabla^{f^{-1}TY}_{\frac{\partial}{\partial x^\alpha}} \left(\frac{\partial f}{\partial s}\right) dx^\alpha , \frac{\partial f}{\partial x^\beta}dx^\beta \right\rangle +$$

$$+ \left\langle \nabla^{f^{-1}TY}_{\frac{\partial}{\partial x^\alpha}} \left(\frac{\partial f}{\partial s}\right) dx^\alpha , \nabla^{f^{-1}}_{\frac{\partial}{\partial x^\beta}} \left(\frac{\partial f}{\partial t}\right) dx^\beta \right\rangle =$$

$$= \left\langle \nabla^{f^{-1}TY}_{\frac{\partial}{\partial x^\alpha}} \nabla_{\frac{\partial}{\partial t}} \left(\frac{\partial f}{\partial s}\right) dx^\alpha , \frac{\partial f}{\partial x^\beta}dx^\beta \right\rangle$$

$$+ \left\langle R^N \left(\frac{\partial f}{\partial x^\alpha}dx^\alpha , \frac{\partial f}{\partial t}\right)\frac{\partial f}{\partial s} , \frac{\partial f}{\partial x^\beta}dx^\beta \right\rangle$$

$$+ \left\langle \nabla^{f^{-1}TY}_{\frac{\partial}{\partial x^\alpha}} v\, dx^\alpha , \nabla^{f^{-1}TY}_{\frac{\partial}{\partial x^\beta}} w\, dx^\beta \right\rangle ,$$

Now

$$\int_X \left\langle \nabla^{f^{-1}TY}_{\frac{\partial}{\partial x^\alpha}} \nabla_{\frac{\partial}{\partial t}} \frac{\partial f}{\partial s} dx^\alpha , \frac{\partial f}{\partial x^\beta}dx^\beta \right\rangle dX$$

$$= \int \frac{\partial}{\partial x^\alpha}\left(\gamma^{\alpha\beta}\left\langle \nabla_{\frac{\partial}{\partial t}} \frac{\partial f}{\partial s} , \frac{\partial f}{\partial x^\beta}\right\rangle_{f^{-1}TY} \sqrt{\gamma}\right) dx^1 \ldots dx^n$$

$$+ \quad \int \left\langle \nabla_{\frac{\partial}{\partial t}} \frac{\partial f}{\partial s} dx^{\alpha} , \nabla_{\frac{\partial}{\partial x^{\alpha}}} \frac{\partial f}{\partial x^{\beta}} dx^{\beta} \right\rangle$$

$$= \int \left\langle \nabla_{\frac{\partial}{\partial t}} \frac{\partial f}{\partial s} , \gamma^{\alpha\beta} \nabla_{\frac{\partial}{\partial x^{\alpha}}} \frac{\partial f}{\partial x^{\beta}} \right\rangle_{f^{-1}TY}$$

by Stokes' Theorem

$$= 0 \text{ , since } \gamma^{\alpha\beta} \nabla_{\frac{\partial}{\partial x^{\alpha}}} \frac{\partial f}{\partial x^{\beta}} = \text{trace } \nabla\, df = 0 \text{ , as } f \text{ is harmonic.}$$

Thus

$$H_f(v,w) = \int_X \gamma^{\alpha\beta} \left\langle \nabla^{f^{-1}TY}_{\frac{\partial}{\partial x^{\alpha}}} v , \nabla^{f^{-1}TY}_{\frac{\partial}{\partial x^{\beta}}} w \right\rangle_{f^{-1}TY}$$

$$- \int_X \gamma^{\alpha\beta} \left\langle R^N\left(\frac{\partial f}{\partial x^{\alpha}} , v\right) \frac{\partial f}{\partial x^{\beta}} , w \right\rangle_{f^{-1}TY}$$

$$= \int_X \left\langle \nabla^{f^{-1}TY} v , \nabla^{f^{-1}TY} w \right\rangle_{f^{-1}TY}$$

$$- \int_X \text{trace}_X \left\langle R^N(df,v)\, df , w \right\rangle_{f^{-1}TY}$$

For the preceding calculations cf. also [EL 4] .
We now want to look at the definition of harmonic maps from a somewhat different point of view. By the famous embedding theorem of Nash ([Na]), Y can be isometrically embedded in some Euclidean space $\mathbb{R}^l$.
We define the Sobolev space

$$W^1_2(X,Y) = \{f \in W^1_2(X,\mathbb{R}^l) \ : \ f(x) \in Y \ \text{ a.e.}\}$$

Since $W^1_2(X,\mathbb{R}^l) = H^1_2(X,\mathbb{R}^l)$ by a well-known theorem of Meyers and Serrin

(cf. [MS] ; we can assume X to be a compact manifold (possibly with boundary), since we always can localize the problem in the domain. Namely, if f is a critical point of E on X , then it is also critical on any subdomain) every element in $W_2^1(X,Y)$ can be approximated with respect to the W_2^1 norm by smooth mappings, namely from $C^\infty(X,\mathbb{R}^1)$, although the corresponding equality $W_2^1(X,Y) = H_2^1(X,Y)$ does not hold in general, cf. [SU 2]. In particular, if we compose an element from W_2^1 (X,Y) with a smooth mapping, we can apply a chain rule.
In this Sobolev space, we can still define the energy functional by

$$E(f) = \frac{1}{2}\int |df(x)|^2 \, dX(x)$$

and look for critical points of E in $W_2^1(X,Y)$.
Assume that $f \in W_2^1(X,Y)$ is a critical point of E which maps X into a compact part Y_0 of Y . Y_0 has a uniform neighborhood in $\mathbb{R}^1$ on which the projection π , mapping a point in $\mathbb{R}^1$ to the closest point in Y , is smooth.
Thus, if $\varphi: X \to \mathbb{R}^1$ is smooth and t is sufficiently small, $(f + t\varphi)(x)$ lies in this neighborhood for a. a. $x \in X$. Since f is critical, we get if $\varphi|\partial X = 0$

$$0 = \frac{\partial}{\partial t} E(\pi(f + t\varphi))|t=0$$

$$= \int_X \left\langle D^2\pi(f) \cdot \varphi D_\alpha f , d\pi(f) D_\alpha f \right\rangle dX$$

$$+ \int_X \left\langle d\pi(f) D_\alpha\varphi , d\pi(f) D_\alpha f \right\rangle dX$$

applying the chain rule,
where $D_\alpha f = e_\alpha(f)$ and e_α is a moving orthonormal frame on X , $\alpha = 1,\ldots n$.

$$= \int_X < D^2\pi(f) \cdot \varphi D_\alpha f , d\pi(f) D_\alpha f > dX +$$

$$+ \int_X < D_\alpha\varphi , d\pi(f) D_\alpha f > dX$$

since π is a projection

$$= \int_X < D^2\pi(f) \cdot \varphi D_\alpha f , D_\alpha f > dX +$$

$$+ \int_X < D_\alpha \varphi , D_\alpha f > dX$$

since $\pi \circ f = f$ and consequently $d\pi \cdot D_\alpha f = D_\alpha f$ by the chain rule. Thus, f is a weak solution of

$$O = \Delta f - D^2\pi(f)(df, df), \tag{1.3.3}$$

where Δ is the Laplace - Beltrami operator on X (cf. [SU 1] for somewhat different calculations). (1.3.1) and (1.3.3) are equivalent, since they both are the Euler - Lagrange equations of the energy functional E. The point of view leading to (1.3.3) was different, however. Here, the energy was minimized among all maps $u: X \to \mathbb{R}^1$ of class $H_2^1 \cap L^\infty(X,\mathbb{R}^1)$ satisfying a nonlinear constraint $u(x) \in Y_0$ (for almost all $x \in X$).

Finally, if we have a harmonic map $u : \Sigma_1 \to \Sigma_2$ between surfaces, and on this surfaces, we have conformal metrics

$$\sigma^2 \, dz \, d\bar{z} = \sigma^2(dx^2 + dy^2) \qquad (z = x + iy)$$

and

$$\rho^2 \, du \, d\bar{u} = \rho^2(du_1^2 + du_2^2) \qquad (u = u_1 + iu_2)$$

then the Laplace - Beltrami operator on Σ_1 is given by $\frac{1}{\sigma^2}\frac{\partial}{\partial z}\frac{\partial}{\partial \bar{z}}$, and (1.3.1) in these coordinates takes the form

$$\frac{1}{\sigma^2} u_{z\bar{z}} + \frac{1}{\sigma^2}\frac{2\rho_u}{\rho} u_z u_{\bar{z}} = 0 , \tag{1.3.4}$$

where a subscript denotes a partial derivative. [1)]
From (1.3.4) we see, that in the case where the domain is a surface, the harmonicity of u depends only on the conformal structure, but not on the particular metric of Σ_1, since we can simply multiply (1.3.4) by σ^2.
The harmonicity of u does depend, however, on the image metric, unless u is holomorphic or antiholomorphic, i.e. $u_{\bar{z}} \equiv 0$ or $u_z \equiv 0$.

We note the following lemma

---

1) Note that we have identified u with its coordinate representation.

Lemma 1.1: If $u : \Sigma_1 \to \Sigma_2$ is a harmonic map between surfaces, then

$$\phi = (|u_x|^2 - |u_y|^2 - 2i \langle u_x, u_y \rangle) dz^2 \qquad (z = x + iy)$$
$$= 4\rho^2 u_z \bar{u}_z \, dz^2$$

is a holomorphic quadratic differential.

Proof: Multiplying (1.3.4) by the conformal factor σ^2, we obtain

$$\tilde{\tau}(u) := u_{z\bar{z}} + \frac{2\rho_u}{\rho} u_z u_{\bar{z}} = 0$$

Thus,

$$\phi_{\bar{z}} = 2\rho\rho_u u_{\bar{z}} u_z \bar{u}_z + 2\rho\rho_{\bar{u}} \bar{u}_{\bar{z}} u_z \bar{u}_z + \rho^2 u_{z\bar{z}} \bar{u}_z + \rho^2 u_z \bar{u}_{z\bar{z}}$$
$$= \rho^2(\bar{u}_z \tilde{\tau}(u) + u_z \bar{\tilde{\tau}}(u)) = 0$$

q.e.d.

We also observe, that if ϕ is holomorphic then $\tau(u) = 0$ with the possible exception of points where $|\bar{u}_z| = |u_z|$, i.e. where the Jacobian $|u_z|^2 - |u_{\bar{z}}|^2$ vanishes. This was actually used by Gerstenhaber and Rauch [GR] as a definition of harmonic maps between surfaces.
We note moreover, that ϕ is just the (2,0) part of the differential form $u^*(4\rho^2(u)du d\bar{u})$, i.e. the pull-back of the image metric under u.
Finally, of course $\phi \equiv 0$ if and only if u is conformal.
Therefore, Lemma 1.1 together with the Theorem of Liouville, which implies that $\phi \equiv 0$ is the only holomorphic quadratic differential on S^2, shows that any harmonic map from S^2 is conformal or anticonformal. We shall see a different proof of this fact in Cor. 12.1.

1.4. Mathematical problems arising from the concept of harmonic maps

From 1.3, one sees that new mathematical difficulties arise compared to the case of geodesics. Here, critical points lead to systems of nonlinear partial differential equations, while geodesics lead only to systems of ordinary differential equations. The natural space to look for critical points of E is the Sobolev space $W_2^1(X,Y) \cap L^\infty(X,Y)$, since the equations for weak solutions of (1.3.1), namely

$$(1.4.1) \qquad 0 = \int \gamma^{\alpha\beta} \left(\frac{\partial f^i}{\partial x^\alpha} \frac{\partial \varphi^i}{\partial x^\beta} - \Gamma^i_{jk} \frac{\partial f^j}{\partial x^\alpha} \frac{\partial f^k}{\partial x^\beta} \varphi^i\right) dx$$

make sense only for test functions $\varphi \in \overset{\circ}{W}{}^1_2(X,\mathbb{R}^N) \cap L^\infty(X,\mathbb{R}^N)$.

From an analytical point of view, it is not surprising that the equations (1.3.1) turned out to be rather difficult to handle, since the nonlinearity is quadratic in the gradient of the solution. Such systems may have nonsmooth weak solutions. This phenomenon can even occur in the present situation. Namely, mapping the unit ball D^n of dimensions $n \geqq 3$ onto its boundary via radial projection, can be interpreted as a weakly harmonic map (i.e. a solution of (1.4.1)) $f: D^n \to S^{n-1}$, cf. [HKW 3] .
Thus, although harmonic maps had been conceived for some time, what got the subject really off the ground, was the paper [ES] by Eells and Sampson, in which the existence of a harmonic map between compact manifolds was proved for the case of nonpositive curvature of the image. Although their proof can be somewhat simplified and freed from some technical assumptions, using subsequent ideas of Hartman [Ht],their result is still basic and has admitted improvements only in special cases.
We shall mention some mathematical applications of harmonic maps in the following chapters, but we feel that the range of possible applications has by no means been yet fully exploited.

1.5. Physical significance

Harmonic maps into spheres or complex projective spaces have also acquired some physical interest since they turned out to be solutions of the nonlinear O(N) σ-models. For more details, we refer to [Mi].

1.6. Some remarks about notation and terminology

If we speak of a surface or a manifold, we usually mean a surface or a manifold together with a given Riemannian metric, unless explicitly stated otherwise (as in Chapter 12). We also note that a Riemannian metric on a surface represents a conformal structure, to which we occasionally refer as the underlying conformal or complex structure. With respect to such structures, we can speak of (anti)holomorphic and (anti)conformal maps.
With the exception of Chapter 3, all manifolds are supposed to be of class C^3. Thus, in particular their curvature tensor is well defined everywhere and continuous. All the results, unless stated otherwise explicitly, hold for manifolds of class C^3. If in some proof, higher regularity is needed, then the C^3 - case follows by straightforward approximation arguments, and we shall not mention it.
Furthermore, manifolds are usually assumed to be compact, or at least complete.

The metric tensor of the domain of a map will be denoted by $(\gamma_{\alpha\beta})$, and $(\gamma^{\alpha\beta}) = (\gamma_{\alpha\beta})^{-1}$, $\gamma = \det (\gamma_{\alpha\beta})$, and the metric tensor of the image manifold by (g_{ij}) and its Christoffel symbols by Γ^i_{jk} . Greek minuscules as indices always refer to the domain, and latin ones to the image. We shall use the Einstein summation convention.
$T_x X$ is the tangent space of the manifold X at the point x .
Sometimes we shall use local coordinates, while at other occasions we shall prefer a more intrinsic calculus, like covariant derivatives, orthonormal frames, complex notations, etc., whatever is most convenient.

If U is a subset of a manifold, we denote by $\overset{\circ}{U}$ its interior and by $\overline{U}$ its closure.
D denotes the closed unit disc in the plane, i.e.

$$D: = \{(x,y) \in \mathbb{R}^2 : x^2 + y^2 \leq 1\} .$$

On a manifold X , we denote a geodesic ball with center $p \in X$ and radius R by B(p,R), i.e.

$$B(p,R): = \{q \in X : d(p,q) \leq R\} ,$$

where d is the distance function on X.
Upper and lower bounds for the sectional curvature K of a manifold are often denoted by κ^2 and $-\omega^2$, i.e.

$$-\omega^2 \leq K \leq \kappa^2 .$$

This notation avoids square roots. It differs, however, from the terminology in some of the papers frequently referred to in the present book.
The functional determinant of a map u at a point x is denoted by J(u)(x) or simply J(x), if the map u is understood from the context.
We shall use the standard mapping spaces $C^{k,\alpha}(X,Y)$ of k times differentiable maps, the k - th derivatives of which satisfy a Hölder condition with exponent α . Here, the Hölder exponent α is always taken from the open unit interval (0,1). If not stated otherwise, results involving $C^{k,\alpha}$ spaces hold for all $\alpha \in (0,1)$. Whether the Sobolev space $H^1_2(X,Y)$ will be defined with the help of a coordinate chart or with the help of an embedding of Y into some Euclidean space, will always be clear from the context.
We try to give selfcontained and complete proofs of most twodimensional results. We assume, however, basic facts from Riemannian geometry as

well as the theory of linear elliptic equations and systems. Good references are [Bl] and [GKM] for the geometry and [GT] for the results about partial differential equations.

2. Geometric considerations

2.1. Convexity, existence of geodesic arcs, and conjugate points

We start with some elementary considerations. The proof of the following lemma, an application of the Theorem of Arzela-Ascoli, is well known.

Lemma 2.1: Suppose M is a compact surface, possibly with boundary. If the boundary γ is nonvoid, it is assumed to be convex. Given two points p and q in M, every homotopy class of arcs between them contains a shortest arc, and this arc is geodesic.

Here, a Lipschitz curve γ is called convex w.r.t.M, if through every point $q \in \gamma$ there goes a geodesic arc which is disjoint to the interior of M in a neighborhood of q.

Prop. 2.1: Under the assumptions of Lemma 2.1, assume that there are two distinct homotopic geodesic arcs joining p and q. Then each of the points p and q has a conjugate point in M, and this point is conjugate to p or q, resp., with respect to a geodesic arc which is the shortest connection in its homotopy class.

Proof: We denote the two geodesic arcs by γ_1 and γ_2. We can assume w.l.o.g. that γ_1 and γ_2 are shortest connections in their homotopy class between p and q, since otherwise, starting e.g. from p and moving on γ_1, we would find a point q_1 which would either be conjugate to p or would have a connection to p in the same homotopy class and of equal length as the segment of γ_1 between p and q_1. (At this point, for the existence of such a connection, we have to use Lemma 2.1 and therefore the convexity of γ). Since γ_1 and γ_2 are homotopic and distinct, because we could assume that they are shortest connections, they bound a set B of the topological type of the disc.

We now look at a geodesic line emanating from p into B. As γ_1 and γ_2 are shortest, this line has to cease somewhere in B to be shortest connection to p. Repeating the argument, if we have not yet found the desired conjugate point, we get a nested sequence of geodesic two-angles, i.e. configurations consisting of two homotopic geodesic arcs of equal length which furthermore are shortest possible in their homotopy class. In the limit, this construction has to yield a geodesic arc covered twice. The endpoint q_2 therefore is conjugate to p , and furthermore, the geodesic arc is the shortest connection in its homotopy class from p to q_2 .

q.e.d.

For the preceding argument, cf. [Bl], p. 231.

2.2. Convexity of the squared distance function

For a function h on a Riemannian manifold, its second fundamental form at a point q is defined via

$$D^2h(q)(Y,Z) = \langle D_Y \operatorname{grad} h(q), Z \rangle ,$$

where Y and Z are tangent vectors at q and D_Y denotes covariant differentiation in the direction Y .

Lemma 2.2: Suppose $B(p,r) = \{q \in M: d(p,q) \le r\}$ is a geodesic ball in a Riemannian manifold M which is disjoint to the cut locus of its center. For the sectional curvature K on B(p,r) we assume

$$-\omega^2 \le K \le \kappa^2 ,$$

and

$$r \le \frac{\pi}{2\kappa} .$$

We define $r(x) := d(x,p)$, $f(x) = \frac{1}{2} d(x,p)^2$.
Then $f \in C^2$, and

a) $|\operatorname{grad} f(x)| = r(x)$

b) $\kappa r(x) \cdot \operatorname{ctg} \kappa r(x) \cdot |Y|^2 \le D^2 f(Y,Y) \le \omega r(x) \coth \omega r(x) \cdot |Y|^2$

Proof: a) follows from $\operatorname{grad} f(x) = -\exp_x^{-1}(p)$.
If q(t) is a curve in B(p,r) with $q(0) = x$ and $\dot{q}(0) = Y$, putting

$$c(s,t) = \exp_{q(t)} (s \exp_{q(t)}^{-1} p),$$

we have

$$\operatorname{grad} f(x) = - \frac{\partial}{\partial s} c(s,t)_{|s = 0}$$

Thus

$$D_Y \text{ grad } f(x) = -\frac{D}{\partial t}\frac{\partial}{\partial s} c(s,t)_{|s=0}$$
$$= -\frac{D}{\partial s}\frac{\partial}{\partial t} c(s,t) .$$

For each t, $J_t(s) = \frac{\partial}{\partial t} c(s,t)$ is the Jacobi field along the geodesic arc $c(\cdot,t)$ with $J_t(0) = \dot{q}(t)$ and $J_t(1) = 0$. Therefore

$$D_Y \text{ grad } f(x) = D_{J(0)} \text{ grad } f(x) = -\dot{J}(0).$$

The Jacobi field estimates (A 4.1) and (A 5.1) from [K2] now imply b). Note that these estimates may also be applied to tangential Jacobi fields, since we assume bounds $-\omega^2 \le K \le \kappa^2$.

2.3. Uniqueness of geodesic arcs in convex discs

<u>Lemma 2.3:</u> <u>Suppose</u> $B(p,R) := \{q \in \Sigma: d(p,q) \le R\}$, <u>where</u> Σ <u>is a surface, is topologically a disc for some</u> $R < \frac{\pi}{\kappa}$ $(K \le \kappa^2)$. <u>Then</u> $\exp_p \{v: |v| = r\} = \partial B(p,r)$ <u>for all</u> $r \le R$, <u>where</u> $\exp_p : T_p\Sigma \to \Sigma$ <u>is the exponential map. Furthermore</u>, $\partial B(p,r)$ <u>is convex, if</u> $r \le \frac{\pi}{2\kappa}$.

<u>Proof:</u> Clearly, $\partial B(p,r) \subseteq \exp_p \{v: |v| = r\} \subseteq B(p,r)$.
We assume now that

$$(2.3.1) \qquad \exp_p \{v: |v| = r\} \cap \overset{\circ}{B}(p,r) \neq \emptyset$$

$\exp_p$ is a local diffeomorphism on $\{v: |v| < \frac{\pi}{\kappa}\}$ by the comparison theorem of Morse - Schoenberg (cf. [GKM], p. 176), and therefore $\exp_p \{v: |v| = r\}$ is an immersed smooth curve for $r < \frac{\pi}{\kappa}$.
Since $\exp_p \{v: |v| = r\}$ is compact, we can find some $q \in \exp_p \{v: |v| = r\}$ with minimal distance to p. Consequently, the shortest geodesic γ from p to q is orthogonal to $\exp_p \{v: |v| = r\}$ at q and has length $< r$. On the other hand, $q = \exp_p w$, $|w| = 0$, and the geodesic $\gamma' = \exp_p tw$, $t \in [0,1]$, is also orthogonal to $\exp_p \{v: |v| = r\}$ at q and different from γ, since its length is precisely r. Thus, γ and γ' have an angle of π at q and match together to a geodesic loop with corner at p. It is not difficult to see that every point inside this geodesic loop can be joined to p by a shortest geodesic, in spite of the fact that this loop might not be convex at p. Thus, we can carry over the argument of Prop. 2.1 to assert the existence of a point p'

inside this loop which is conjugate to p w.r.t. a shortest geodesic γ''. Since $p' \in B(p,r)$ and $r < \frac{\pi}{\kappa}$, this is in contradiction to the comparison theorem of Morse - Schoenberg. This proves the first claim. Furthermore, since $\exp_p$ has maximal rank on $\{v \in T_p\Sigma : |v| < \frac{\pi}{\kappa}\}$, as noted above, we infer that every $v \in T_p\Sigma$ with $|v| = r$ has a neighborhood V which is mapped under $\exp_p$ injectively onto its image (cf. [K1], p. 108f.). From this, we easily see that we may apply the estimate of Lemma 2.2b. Therefore, if $r \le \frac{\pi}{2\kappa}$, then f is a convex function on $B(p,r)$, and consequently, $\partial B(p,r) = \exp_p \{v: |v| = r\}$ is convex as a level set of a convex function.

q.e.d.

Thm. 2.1: Suppose now, that $B(p,r)$ is a geodesic disc on a surface, and $r < \frac{\pi}{2\kappa}$ $(K \le \kappa^2)$. Then each pair of points q_1, $q_2 \in B(p,r)$ can be joined by a unique geodesic arc in $B(p,r)$, and this arc is free of conjugate points.

Proof: By virtue of Lemma 2.3, we could apply Prop. 2.1, if there would exist two geodesic arcs joining q_1 and q_2. Consequently, we would find a point q_3 conjugate to q_1 w.r.t. a shortest geodesic arc, i.e. an arc of length $\le 2r < \frac{\pi}{\kappa}$. This would contradict the Morse - Schoenberg Theorem (cf. [GKM], p. 176).

q.e.d.

2.4. Remark: The higher dimensional analogue of 2.3.

Thm. 2.1 pertains to arbitrary dimension, if we suppose a priori that $B(p,r)$ is disjoint to the cut locus of p. The proof is different, however, since the argument of Prop. 2.1 is clearly restricted to two dimensions. Cf [J2] for a proof in the general case.

2.5. Curvature of parallel curves

Let now c be a C^2 curve without double points on the surface M. One can measure the distance to c on M, and since locally c divides M, one can assign a negative sign to this distance function on one side of c, thus obtaining a C^2 function h on a neighborhood of c. If q is a point in this neighborhood, c_1 the curve through q which is parallel to c, then grad h(q) is the unit tangent vector of the geodesic through q which is orthogonal to c_1. If $Y, Z \in T_qM$, we have

(2.5.1) $$D^2h(q)(Y,Z) = \langle D_Y \operatorname{grad} h(q), Z\rangle = 0$$

in case Y or Z is orthogonal to $\dot{c}_1$
and

(2.5.2) $$D^2h(q)(Y,Y) = \kappa_g(c_1;q)|Y|^2 ,$$

in case Y is tangential to c_1 , where κ_g denotes geodesic curvature.

Lemma 2.4: Suppose c is geodesic, $p \in M$ is not further away from c than $\frac{\pi}{2\Lambda}$ ($\Lambda^2 := \max |K|$) or a cut point. Then

(2.5.3) $$|D^2h(q)(Y,Y)| \leq \Lambda|\tan(\Lambda h(q))| \cdot |Y|^2 \quad \text{for } Y \in T_pM .$$

Proof: For the geodesic curvature $\kappa_g(h)$ of the parallel curves of c at distance h , we have the differential equation

(2.5.4) $$\kappa_g'(h) = \kappa_g^2(h) + K(h) ,$$

and thus

$$|\tfrac{1}{\Lambda} \text{arctg } (\tfrac{1}{\Lambda}\kappa_g)'| \leq 1 ,$$

and since by assumption $\kappa_g(0) = 0$,

(2.5.5) $$|\kappa_g(h)| \leq \Lambda|\tan \Lambda h| ,$$

and (2.5.3) now follows from (2.5.1) and (2.5.2).

2.6. Local coordinates with curvature controlled Christoffel symbols

In this section, we want to introduce coordinates with curvature controlled Christoffel symbols in a neighborhood of a point $q \in B(p,M)$, without using any information of the geometry outside $B(p,M)$. The following construction is taken from [JK1] and works in any dimension. We shall restrict ourselves to the twodimensional case, however, for simplicity. We suppose again that $B(p,M)$ is a disc with $M < \frac{\pi}{2\kappa}$, in order to be able to apply Thm. 2.1.
In case $d(p,q) \leq \frac{1}{2}M$, we take an arbitrary orthonormal base e_1 , e_2 of T_qY .
If $d(p,q) > \frac{1}{2}M$, we choose e_1 and e_2 in such a way that their angle with the geodesic arc from q to p is $\frac{\pi}{4}$. We now want to show that the geodesics $\exp_p(t\cdot e_i)$ stay inside $B(p,M)$ for $t \leq t_0$, where $t_0 > 0$ can

be estimated from below in terms of ω and M.

Indeed, by the Rauch-Toponogow Comparison Theorem (cf. [GKM], p. 194f),

$$d(p, \exp_q t\cdot e_i) \le d^\omega(\tilde{p}, \exp_{\tilde{q}} t\cdot \tilde{e}_i) ,$$

where the right hand side is the distance in the comparison triangle in the plane of constant curvature $-\omega^2$, with $d^\omega(\tilde{p},\tilde{q}) = d(p,q)$, $\tilde{e}_i$ having again an angle of $\frac{\pi}{4}$ with the geodesic form $\tilde{q}$ to $\tilde{p}$. Consequently

$$\begin{aligned} \cosh(\omega d(p, \exp_q te_i)) &\le \cosh \omega t \cdot \cosh(\omega d(p,q)) - \\ &- \tfrac{1}{2} \sinh \omega t \cdot \sinh(\omega d(p,q)) \le \\ &\le \cosh \omega t \cdot \sinh \omega M - \\ &- \tfrac{1}{2} \sinh \omega t \cdot \sinh \omega M , \end{aligned}$$

if $t \le \frac{1}{2} M$

$$\le \cosh \omega M ,$$

if $t \le \bar{t}$, say.

Then, for $t \le t_o = \min (\bar{t}, \frac{1}{2} M)$, $d(p, \exp_q te_i) \le M$, and consequently the geodesics $\exp_q te_i$ stay inside $B(p,M)$ for $t \le t_o$.

Lemma 2.5 ([JK1]): In a neighborhood $B(q,\tau) \cap B(p,M)$ of $q \in B(p,M)$, we can define local coordinates for which the Christoffel symbols are bounded in absolute value and $\tau > 0$ is bounded from below, both in terms of ω, κ, M only, via

$$h_i(s) := \frac{1}{2t_o} (d^2(s, \exp_q t_o e_i) - d^2(s,q)) .$$

Proof: By Lemma 2.2,

$$(2.6.1) \qquad |D^2 h_i(s)| \le \frac{\omega M}{t_o} \coth \omega \frac{M}{2}$$

if $d(s,q) \le \frac{1}{2} M$, and

$$(2.6.2) \qquad dh_{|q} \text{ is an isometry },$$

where $h = (h_1 , h_2): B(p,M) \to \mathbb{R}^2$.

This easily implies a lower bound τ for the radius of the set on which h is injective.

Furthermore, the Christoffel symbols are given by $D^2 h$, namely

$$(2.6.3) \qquad D^2_{X,Y}h = \langle D_X \operatorname{grad} h, Y\rangle = d_{dh\cdot X}dh\cdot Y - dh\cdot D_X Y$$
$$= d_{dh\cdot X}dh\cdot Y - dh\cdot d_X Y - dh\cdot \Gamma(X,Y)$$
$$= -dh\cdot \Gamma(X,Y) \ ,$$

since dh is linear.
Thus, Lemma 2.2 also implies the bound on the Christoffel symbols.

q.e.d.

Remark: In chapter 10, we shall obtain coordinates with even better regularity properties, namely harmonic coordinates.

3. Conformal mappings

3.1. Statement of Thm. 3.1 concerning conformal representations of compact surfaces homeomorphic to plane domain

Theorem 3.1: Suppose S is a surface with boundary, homeomorphic to a plane domain G bounded by k circles via a chart $\psi: \overline{G} \to S$. Suppose the coefficients of the metric tensor of S can be defined in this chart by bounded measurable functions g_{ij} with $g_{11}\,g_{22} - g_{12}^2 \geq \lambda > 0$ in G. Then S admits a conformal representation $\tau \in H^1_2 \cap C^\alpha(\overline{B},\overline{G})$, where B is a plane domain bounded by k circles and τ satisfies almost everywhere the conformality relations

$$|\tau_x|^2 = |\tau_y|^2 \quad , \quad \tau_x \cdot \tau_y = 0$$

(here (x,y) denote the coordinates of points in $\overline{B}$, and norms and products are taken with respect to the metric of S.)
τ can be normalized by a three point condition, namely three points on one of the boundary curves of S can be made to correspond, respectively, to three given points on the outer boundary of B which can be taken as the unit circle, or by fixing the image of an interior point. Furthermore, concerning higher regularity, τ is as regular as S, i.e. if S is of class $C^{k,\alpha}$ $(k \in \mathbb{N},\ 0 < \alpha < 1)$ or C^∞, then also $\tau \in C^{k,\alpha}(\overline{B})$ or $C^\infty(\overline{B})$, respectively. In particular, if S is at least $C^{1,\alpha}$, then the conformality relations are satisfied everywhere, and τ is a diffeomorphism.

Remark: For an explicit interior $C^{2,\alpha}$ estimate for τ, cf. Cor. 10.2.

For higher regularity, see also Cor. 10.3.

In his monograph [M3], Morrey states a corresponding theorem for surfaces of class $C^{1,\alpha}$ in Euclidean 3 - space (the latter restriction is obviously superfluous and stems only from the fact that Morrey wanted to apply this theorem in his investigation of the Plateau problem for minimal surfaces in Euclidean 3 - space).
In this case, a local version was proved by Lichtenstein [Li], and this local version together with the uniformization theorem yields the one-contour case (k = 1).
In the general case, a local result was proved by Lavrent'ev [Lv] (for continuous metrics) and Morrey [M1] (for measurable metrics). (cf. [BJS] for more details), and one can again combine this result with the uniformization theorem in case $k = 1$.
In [M3], however, Morrey did not appeal to the uniformization theorem or a local result and tried to prove the global theorem directly by a variational method.
In his proof, Morrey constructed an energy minimizing sequence from circular domains onto G , in order to obtain a conformal limit map τ . His proof of the equicontinuity of the minimizing sequence, however, is not correct, since he claims implicitly on p. 368, that, if the oscillation of a continuous H_2^1 map on a disc is bounded below by ε , then the length of the image of the boundary curve of this disc is at least $\varepsilon/2$. It is easy to construct counterexamples to this assertion.

Also, it is not clear to me, why $\psi\circ\tau: B \to S$ represents the same Fréchet surface as $\psi: G \to S$ which is needed in his proof to apply Lemma 9.3.7, if we do not know a priori that τ is a uniform limit of diffeomorphisms.
(Finally, Morrey did not prove that τ is a diffeomorphism (in case $S \in C^{1,\alpha}$) or at least a homeomorphism. We shall come back to this question in sections 3.5 and 3.6.)
In the sequel, I shall show, how one can easily overcome these difficulties and even simplify Morrey's proof by choosing the minimizing sequence in a restricted, yet large enough class to show the conformality of the limit map. (A similar idea shall also turn out to be of crucial importance in later chapters).
Since some of the arguments can be taken over directly from [M3], we shall only sketch these.

3.2. The Courant - Lebesgue Lemma

In the proof of Theorem 3.1, we shall make use of a well - known lemma due to Lebesgue and Courant (cf. [Co], p. 101).

We state it here in a more general form than we need for this chapter, since this form will be needed in subsequent chapters.
Suppose Ω is an open subset of some twodimensional Riemannian manifold Σ of class C^3, while S satisfies the same assumptions as in Thm. 3.1.

Lemma 3.1: Let $u \in H_2^1(\Omega, S)$, $E(u) \le D$, $x_0 \in \Sigma$, $-\lambda^2$ a lower bound for the curvature K of Σ, $\delta < \min(1, i(\Sigma)^2, 1/\lambda^2)$.
Then there exists some $r \in (\delta, \sqrt{\delta})$ for which $u|\partial B(x_0, r) \cap \bar{\Omega}$ is absolutely continuous and

$$d(u(x_1), u(x_2)) \le 4\pi \cdot D^{1/2} \cdot (\log 1/\delta)^{-1/2}$$

for all $x_1, x_2 \in \partial B(x_0, r) \cap \bar{\Omega}$.

Proof: We introduce polar coordinates on $B(x_0, r)$, i.e. $ds^2 = dr^2 + G^2(r,\theta)d\theta^2$.
Since $K = -\frac{G_{rr}}{G}$ (cf. [Bl], p. 153), we infer

$$G(r,\theta) \le 1/\lambda \sinh \lambda r . \tag{3.2.1}$$

Now for $x_1, x_2 \in \partial B(x_0, r)$ and almost all r, since u is a Sobolev function $u|\partial B(x_0, r)$ is absolutely continuous and

$$\begin{aligned} d(u(x_1), u(x_2)) &\le \int_0^{2\pi} |u_\theta(x)| d\theta \\ &\le 2\pi \left(\int_0^{2\pi} |u_\theta|^2 d\theta \right)^{1/2} \end{aligned} \tag{3.2.2}$$

where we assumed w.l.o.g. $B(x_0, r) \subset \Omega$.
The Dirichlet integral of u on $B(x_0, r)$ is

$$E(u; B(x_0, r)) = 1/2 \int_{B(x_0, r)} (|u_r|^2 + \frac{1}{G^2} |u_\theta|^2) G \, dr \, d\theta$$

Thus, we can find some $r \in (\delta, \sqrt{\delta})$ with

$$\int_0^{2\pi} |u_\theta(r,\theta)|^2 d\theta \le \frac{2D}{\int_\delta^{\sqrt{\delta}} \frac{1}{G(\rho,\theta)} d\rho} \le \frac{2D}{\log 1/\delta} \tag{3.2.3}$$

since for $r \le \sqrt{\delta} \le 1/\lambda$, $G(r,\theta) \le 2r$ by (3.2.1).
The lemma follows from (3.2.2) and (3.2.3).

q.e.d.

3.3. Proof of Theorem 3.1

By assumption, there is a diffeomorphism of some circular domain $\bar{B}$ onto $\bar{G}$. Using elementary Möbius transformations, we can assume that the outer boundary of B is the unit circle and that this diffeomorphism satisfies the three point condition.
We then define $\mathfrak{D}$ as the class of all diffeomorphisms from a circular domain bounded by k circles, the outer boundary of which is the unit circle, onto G , satisfying the three point condition.
We let $\bar{\mathfrak{D}}$ be the class of all maps t: B → G where B is a circular domain of the type considered and t is the weak H_2^1 and uniform limit of maps t_n: B → G from $\mathfrak{D}$.
We now assume for a moment that the g_{ij} are continuous. This assumption will be removed in 3.6. We define the energy of a map ρ: B → G as

$$E(\rho) = \frac{1}{2}\int g_{ij}(\rho(x))D_\alpha\rho^i D_\alpha\rho^j dx \qquad (\alpha = 1,2)$$

We now want to show that we can bound the modulus of continuity of any $\tau \in \bar{\mathfrak{D}}$ in terms of $E(\tau)$ only. Let $E(\tau) \le D$, say.
By the Courant - Lebesgue Lemma 3.1, for every $x_o \in \bar{B}$ and $\delta < 1$, we can find some $r \in (\delta, \sqrt{\delta})$ with

$$(3.3.1) \qquad d(\tau(x_1), \tau(x_2)) \le 2\pi\cdot D^{1/2}\cdot(\log 1/\delta)^{-1/2}$$

for all x_1 , $x_2 \in \partial B(x_o, r) \cap \bar{B}$.

Now $\partial B(x_o, r)$ divides $\bar{B}$ into a "small" and a "large" region, where the small region $B(x_o, r) \cap \bar{B}$ contains at most one of the points p_1 , p_2 , p_3 from the three point condition. Likewise $\tau(\partial B(x_o, r) \cap \bar{B})$ divides S into two regions, the small one containing at most one of the points $\tau(p_1)$, $\tau(p_2)$, $\tau(p_3)$, since τ is a uniform limit of diffeomorphisms. Furthermore, the diameter of the small region goes uniformly to zero as δ goes to zero, since S is compact and its metric is uniformly equivalent to the Euclidean one of G.
Since τ is a uniform limit of diffeomorphisms, the small region in B has to be mapped into the small region in S , and we obtain the desired estimate of the modulus of continuity.
We now choose an energy minimizing sequence τ_n: B_n → G in $\bar{\mathfrak{D}}$, where the B_n are circular domains of the required type. We can again assume $E(\tau_n) \le D$ for some finite D, and hence the τ_n are equicontinuous. This implies that no boundary circle of the domains B_n can collaps to a point or become tangent to another one in the limit as $n \to \infty$. After selection of a subsequence, the B_n then have to converge to some limiting domain B of the required type, and we can assume w.l.o.g. that all τ_n are defined on B .

Thus, a subsequence converges weakly in H_2^1 and uniformly to a limit $\tau \in \bar{\mathfrak{D}}$. Because of the lower semicontinuity of the Dirichlet integral with respect to weak H_2^1 convergence, τ minimizes energy in $\mathfrak{D}$ (Note for this point that, taking $\bar{G}$ as a coordinate domain for S, the energy of τ_n is uniformly equivalent to the Dirichlet integral of τ_n, regarded as a mapping from B_n onto G, because of the compactness of S). Furthermore, τ is continuous.

We now wnat to show, using the standard argument that τ is weakly conformal (cf. [Co], pp. 169 - 178, or [M3], pp. 369 - 372).

To achieve this, we shall compare τ with $\tau \circ \sigma_\lambda$, where $\sigma_\lambda : B_\lambda \to B$ is a family of diffeomorphisms from circular domains B_λ onto B depending smoothly on the real parameter λ, with $\sigma_0 = \mathrm{id}: B \to B$. We observe first that we don't have to require that σ_λ preserves the three point condition nor that the outer boundary of B_λ is the unit circle, since this can be achieved by composition with a conformal map without changing the energy of $\tau \cdot \sigma_\lambda$ and secondly that therefore $\tau \cdot \sigma_\lambda$ are admissible comparison maps, i.e. that again after a possible composition with a conformal map, $\tau \circ \sigma_\lambda \in \bar{\mathfrak{D}}$.

Since τ minimizes E in $\bar{\mathfrak{D}}$, we infer

$$\frac{d}{d\lambda} E(\tau \circ \sigma_\lambda^{-1})_{|\lambda=0} = 0 \qquad (3.3.2)$$

We put

$$E: = g_{ij} \frac{\partial \tau^i}{\partial x} \frac{\partial \tau^j}{\partial x}, \quad F: = g_{ij} \frac{\partial \tau^i}{\partial x} \frac{\partial \tau^j}{\partial y}, \quad G: = g_{ij} \frac{\partial \tau^i}{\partial y} \frac{\partial \tau^j}{\partial y}$$

(note that since $\tau \in H_2^1$, these expressions are defined (only) almost everywhere),

$$\sigma_\lambda : = \xi + i\eta$$

$$\frac{\partial \sigma_\lambda}{\partial \lambda}_{|\lambda=0} = \nu + i\omega .$$

Then

$$E(\tau) = \frac{1}{2} \int_B (E + G)\,dxdy$$

and

$$E(\tau \circ \sigma_\lambda^{-1}) = \frac{1}{2} \int_B \{E(\xi_y^2 + \eta_y^2) - 2F(\xi_x \xi_y + \eta_x \eta_y) + G(\xi_x^2 + \eta_x^2)\} (\xi_x \eta_y - \xi_y \eta_x)^{-1} dxdy ,$$

and (3.3.2) thus implies, noting $\sigma_0(z) = x + iy$,

(3.3.3) $$\int_B (E-G)(\nu_x - \omega_y) + 2F(\nu_y + \omega_x)\,dxdy = 0$$

Putting $\varphi := E - G - 2iF$, (3.3.3) means

$$\mathrm{Re} \int_B \varphi(\nu + i\omega)_{\bar z}\,dxdy = 0$$

Replacing $\nu + i\omega$ by $\omega - i\nu$, we see that the imaginary part likewise vanishes, hence

(3.3.4) $$\int_B \varphi(\nu + i\omega)_{\bar z}\,dxdy = 0$$

First we observe that in (3.3.4) we can insert arbitrary smooth ω and ν with compact support in B , since

(3.3.5) $$\sigma_\lambda(\;) := x + \lambda\nu(z) + i(y + \lambda\omega(z))$$

then is for small $|\lambda|$ a diffeomorphism with the required properties. Hence (3.3.4) implies that φ is holomorphic, i.e.

(3.3.6) $$\varphi_{\bar z} = 0 \qquad \text{in } B .$$

Let $\gamma_1,\ldots,\gamma_k$ be the boundary circles of B , with centers $z_1,\ldots,z_k$, resp. $\gamma_{j,r} := \{z \in B: \mathrm{dist}(z,\gamma_j) = r\}$ $(j \in \{1,\ldots,k\})$. We now look at variations ν and ω which are supported in a neighborhood of γ_j $(j \in \{1,\ldots,k\})$. Integrating (3.3.3) by parts (noting $\varphi \in L_1(B)$), we obtain

(3.3.7) $$\lim_{r\to 0} \int_{\gamma_{j,r}} (E-G)(\nu dy + \omega dx) + 2F(\omega dy - \nu dx) = 0$$

Choosing variations which translate the center z_j , i.e. putting $\nu + i\omega = a_j + ib_j = \text{const.}$ near γ_j (and σ_λ again as in (3.3.5)), we get

(3.3.8) $$\int_{\gamma_{j,r}} \varphi(z)dz = 0 \qquad (r > 0 \text{ small enough})$$

If we apply a homothetic dilation of γ_j instead, i.e. choose $\nu + i\omega = (e_j + if_j)e^{i\theta}$ $(e_j + if_j = \text{const.})$ near γ_j , where θ parametrizes γ_j , we obtain

(3.3.9) $$\int_{\gamma_{j,r}} (z - z_j)\varphi(z)dz = 0 \qquad (r > 0 \text{ small enough})$$

Thus we see that since the boundary circles γ_j were subject to variation, the corresponding periods of φ have to vanish.
This implies that there is an analytic function ψ on B with

$$\psi''(z) = \varphi(z)$$

Since $\varphi \in L_1(B)$, $\psi \in H_1^2(B)$, and hence ψ is continuous on $\overline{B}$.
We now exploit that τ solves a free boundary value problem. We therefore can compose τ with diffeomorphisms rotating γ_j into itself. We thus put

$$\nu + i\omega = \Lambda(\theta) i e^{i\theta} \qquad \text{near } \gamma_j .$$

For simplicity of notation, we assume j = 1 and $\gamma_1 = \{|z| = 1\}$.
(3.3.7) then implies

$$\text{(3.3.10)} \qquad \lim_{r\to 0} \int_{\gamma_{1,r}} \Lambda(\theta) \operatorname{Im} (\psi''((1-r)e^{i\theta})e^{2i\theta})\,d\theta = 0$$

Since $\psi' \in H_1^1(B)$, ψ' has boundary values in $L_1(\partial B)$, and hence we can put r = 0 after an integration by parts; thus

$$0 = \operatorname{Im} \int_{\gamma_1} \psi'(e^{i\theta})(ie^{i\theta}\Lambda'(\theta) - e^{i\theta}\Lambda(\theta))\,d\theta$$

$$= \operatorname{Im} \int_{\gamma_1} \psi'(e^{i\theta}) ie^{i\theta}\Lambda'(\theta)\,d\theta - \operatorname{Im} \int_{\gamma_1} \psi'(e^{i\varphi})e^{i\varphi}(\Lambda(0) + \int_0^{\varphi}\Lambda'(\theta)\,d\theta)\,d\varphi$$

$$= \operatorname{Im} \int_0^{2\pi} \Lambda'(\theta)\,(ie^{i\theta}\psi'(e^{i\theta}) - \int_0^{2\pi} \psi'(e^{i\varphi})e^{i\varphi}d\varphi)\,d\theta$$

$$+ \operatorname{Im} (i\Lambda(0) \int_0^{2\pi} \frac{d}{d\varphi} \psi(e^{i\varphi})\,d\varphi)$$

Since the second term in the last expression vanishes and $\Lambda(\theta)$ was an arbitrary smooth function, we infer from the Lemma of Du Bois - Raymond

$$\text{(3.3.11)} \qquad \operatorname{Im}(iz\psi'(z) - i\psi(z)) = \text{const. on } \gamma_1$$

Therefore, $z\psi'(z) - \psi(z)$ is analytic across γ_1, and thus ψ is smooth along γ_1. Differentiating (3.3.11) w.r.t θ, we obtain on γ_1

$$0 = \operatorname{Im}(iz \frac{d}{dz} (iz\psi' - i\psi)) = -\operatorname{Im} z^2\varphi(z) ,$$

or for the general boundary curve γ_j

(3.3.12) $$0 = \operatorname{Im}\{(z - z_j)^2 \varphi(z)\} \qquad \text{for } z \in \gamma_j .$$

In particular, φ can be analytically continued across ∂B .
Now let $f_j(\theta) = (z - z_j)^2 \varphi(z)$ on γ_j. By (3.3.12), f_j is real on γ_j .
Furthermore, from (3.3.9) and (3.3.8)

(3.3.13) $$\int_0^{2\pi} f_j(\theta)\, d\theta = 0$$

(3.3.14) $$\int_0^{2\pi} f_j(\theta) \cos\theta\, d\theta = \int_0^{2\pi} f_j(\theta) \sin\theta\, d\theta = 0$$

Since f_j is 2π-periodic, (3.3.13) and (3.3.14) imply that f_j and hence φ has at least 4 zeros on γ_j .
Denote the zeros of φ on ∂B by z_s and let

$$B_\rho := B \setminus \bigcup_s (B \cap B(z_s, \rho)) \qquad (\rho > 0)$$

Let ρ be so small that the balls $B(z_s, \rho)$ contain no zeros of φ other than z_s .
The number of zeros of φ inside B_ρ is nonnegative and given by

(3.3.15) $$n = \frac{1}{2\pi i} \int_{\partial B_\rho} d \log \varphi$$

Here, ∂B_ρ is oriented in such a way that B_ρ is to the left.
As $\rho \to 0$, the contribution by $\partial B(z_s, \rho) \cap B$ to the integral in (3.3.15) is $-1/2$ for each z_s , or altogether at least $-2k$, since there are at least $4k$ zeros z_s . (Note that the orientation of ∂B_ρ is such that $B(z_s, \rho)$ is always to the right).
On $\gamma_j \cap B_\rho$, the contribution is given by

$$\frac{1}{2\pi i} \left\{ \int_{\gamma_j \cap B_\rho} d \log ((z - z_j)^2 \varphi(z)) - \int_{\gamma_j \cap B_\rho} d \log (z - z_j)^2 \right\}$$

Since $(z - z_j)^2 \varphi(z)$ is real on γ_j , the first integral tends to zero as $\rho \to 0$, while the second one gives

$$\int_{\gamma_j} d \log (z - z_j)^2 = -2 \qquad \text{for } j = 2, \dots, k$$
$$= 2 \qquad \text{for } j = 1$$

(because of the orientation of ∂B_ρ).
Altogether, we have in (3.3.15), as $\rho \to 0$

$$n = \frac{1}{2\pi i} \int_{\partial B_\rho} d \log \varphi \leq -2k + 2(k-1) - 2 \leq -4$$

which is a contradiction, since n is nonnegative.
The only possibility which avoids this contradiction is that

$$\varphi \equiv 0 ,$$

i.e. that τ is weakly conformal.
Thus

$$|\tau_x|^2 = |\tau_y|^2 \quad \text{and}$$

$$\langle \tau_x , \tau_y \rangle = 0 \tag{3.3.16}$$

almost everywhere in B (so far, these expressions are only known to be in $L^1(B)$).
(If S is parametrized by local coordinates τ^1 , τ^2 with corresponding metric tensor (g_{ij}), the equations (3.3.16) take the form

$$g_{ij}\tau_x^i\tau_x^j = g_{ij}\tau_y^i\tau_y^j$$

$$g_{ij}\tau_x^i\tau_y^j = 0$$

for almost all points $(x,y) \in B \subset \mathbb{R}^2$.)
To repeat, since this point is of crucial importance: Since we need only variations arising from composition with diffeomorphisms of the domain, we can conclude that τ is a solution of the unconstrained variational problem, although we work in an a priori restricted subclass, namely $\overline{\mathfrak{D}}$.

3.4. Proof of Theorem 3.1, continued

For notational convenience, we now call the local coordinates on S $u = \tau^1$, $v = \tau^2$.

We want to exploit that τ is weakly (anti)conformal and the uniform limit of diffeomorphisms in order to show that the Jacobian $u_x v_y - u_y v_x$ of τ has the same sign almost everywhere in B (cf. 9.3.7., l.c.). Here, additional difficulties arise from the fact that τ so far is only known to be of class $C^0 \cap H_2^1$, but these problems can be overcome with the arguments of Lemmata 9.2.4, 9.2.5, l.c.

For the sake of completeness, we provide all details.

Def. 3.1: Suppose G is a plane domain of class C^1, $\varphi \in C^1(G,\mathbb{R}^2)$, $z \notin \varphi(\partial G)$.

Then $m(z,\varphi(\partial G))$ is defined to be the winding number of the curve $\varphi(\partial G)$ w.r.t. z.

If only $\varphi \in C^0(G,\mathbb{R}^2)$, then

$$m(z,\varphi(\partial G)) := \lim_{n\to\infty} m(z,\varphi_n(\partial G))$$

for any sequence $\varphi_n \in C^1(\partial G,\mathbb{R}^2)$ which converges uniformly to φ on ∂G.

That $m(z,\varphi(\partial G))$ is well defined, follows from elementary properties of winding numbers (cf. e.g. [Fe]).

Lemma 3.2: G a plane domain, $\varphi \in C^0 \cap H_2^1(G,\mathbb{R}^2)$.

Then for every $x_0 \in G$, there exists a set $C(x_0)$ with $\mathfrak{H}^1(C(x_0)) = 0$, where $\mathfrak{H}^1$ is 1-dimensional Hausdorff measure, such that for all $R \notin C(x_0)$

$$\int_{B(x_0,R)} J(\varphi)\,dx = \int_{\varphi(B(x_0,R))} m(z,\varphi(\partial B(x_0,R))\,dz$$

if $B(x_0,R) \subset\subset G$.

$$(J(\varphi) := \varphi_x^1\varphi_y^2 - \varphi_y^1\varphi_x^2)$$

Proof: We can find a sequence $\varphi_n \in C^1(D)$, $D \subset\subset G$, converging uniformly and strongly in H_2^1 to φ, so that $\varphi_n \to \varphi$ strongly in $H_2^1(\partial B(x_0,R))$ on $\partial B(x_0,R)$, if $R \notin C(x_0)$, $\mathfrak{H}^1(C(x_0)) = 0$.

Since $H_2^1(\partial B(x_0,R))$ functions are absolutely continuous, and the lengths of $\varphi_n(\partial B(x_o,R))$ and $\varphi(\partial B(x_o,R)$ are uniformly bounded, the twodimensional measure of $\varphi(\partial B(x_o,R)$ vanishes $(R \notin C(x_o))$. Consequently, $z \notin \varphi(\partial B(x_o,R))$ for almost all z, and thus

$$(3.4.1) \qquad m(z,\varphi_n(\partial B(x_o,R)) \to m(z,\varphi(\partial B(x_o,R)) \qquad \text{for these } z .$$

Now

$$\lim_{n\to\infty} \int_{\varphi_n(B(x_o,R))} m(z,\varphi_n(\partial B(x_o,R))dz$$

$$= \lim_{n\to\infty} \int_{B(x_o,R)} J(\varphi_n)dx = \int_{B(x_o,R)} J(\varphi)dx$$

Since $\int_I m(z,\varphi_n(\partial B(x_o,R))dz \leq \left(\frac{\text{meas } I}{\pi}\right)^{1/2}$ length $(\varphi_n(\partial B(x_o,R))$, for any measurable set I, we can integrate (3.4.1), and the result follows.

<u>Lemma 3.3</u>: <u>We suppose that φ_n $(n\in\mathbb{N})$ are diffeomorphisms, converging uniformly and weakly in H_2^1 to φ.</u>
<u>Then $J(\varphi)$ has the same sign almost everywhere.</u>

<u>Proof</u>: Let $B(x_o,R)$, $R \notin C(x_o)$ satisfy the assumptions of Lemma 3.2.

$$\varepsilon_n := \max_{x\in\partial B(x_o,R)} |\varphi_n(x) - \varphi(x)|$$

$$V_n := \{z : d(z,\varphi(\partial B(x_o,R)) > \varepsilon_n\} .$$

For $z \in V_n$, $m(z,\varphi_n(\partial B(x_o,R)) = m(z,\varphi(\partial B(x_o,R))$.
Lemma 3.2 therefore implies

$$(3.4.2) \qquad \lim_{n\to\infty} \int_{\varphi_n^{-1}(V_n)\cap B(x_o,R)} J(\varphi_n) = \int_{B(x_o,R)} J(\varphi) .$$

Since we can assume w.l.o.g., $J(\varphi_n) \geq 0$ in $B(x_o,R)$ for all n, and (3.4.2) holds for almost all discs $B(x_o,R)$, the result follows.
Thus, τ is a weak solution of the corresponding Cauchy - Riemann equations, i.e.

$$(3.4.3) \qquad v_x = -g_{22}^{-1}(g_{12}u_x + k\sqrt{g}\, u_y)$$

$$v_y = g_{22}^{-1}(k\sqrt{g}\, u_x - g_{12}\, u_y)$$

$(g = g_{11}g_{22} - g_{12}^2)$, where $k = \pm 1$ is constant in $\overline{B}$ by Lemma 3.3.
Since this is a first order linear elliptic system, elliptic regularity theory and the usual bootstrap argument imply those assertions in Theorem 3.1, which concern the regularity of τ.
In particular, if S is at least $C^{1,\alpha}$, then τ is a classical solution of (3.4.3) and satisfies the conformality relations (3.3.16) everywhere.

3.5. Proof of Theorem 3.1, continued

In order to show the univalency of τ, we have to prove several lemmata.

Lemma 3.4: Suppose S is of class $C^{2,\alpha}$, i.e. $\psi \in C^{2,\alpha}(\overline{G})$ in the notations of Theorem 3.1. Then τ is a homeomorphism between $\overline{B}$ and S.

Proof: We assume that τ is not a homeomorphism. Then τ is not injective, i.e. there must exist two points z_1, z_2, $z_1 \neq z_2$ with $\tau(z_1) = \tau(z_2)$. We choose a shortest segment γ_n joining $\tau_n(z_1)$ and $\tau_n(z_2)$. Since τ_n is a homeomorphism, $\tilde{\gamma}_n := \tau_n^{-1}(\gamma_n)$ is a curve joining z_1 and z_2.
If $p_{n,\delta}$ is a point on $\partial B(z_1,\delta) \cap \tilde{\gamma}_n$, then for $n \to \infty$ we can find a subsequence of $(p_{n,\delta})$ converging to some point p_δ on $\partial B(z_1,\delta)$. Since the τ_n converge uniformly to τ, we see that $\tau(p_\delta) = \tau(z_1) = \tau(z_2)$.
Thus, a whole continuum is mapped onto the single point $\tau(z_1) = \tau(z_2)$ by τ.
At interior points, we can choose again local coordinates u, v, this time of class $C^{2,\alpha}$. The corresponding metric tensor then is of class C^1, and consequently the corresponding Christoffel symbols can be defined. From (3.4.3) we conclude that u and v are harmonic, e.g.

$$\Delta u + \Gamma_{11}^1(u_x^2 + u_y^2) + 2\Gamma_{12}^1(u_x v_x + u_y v_y) + \Gamma_{22}^1(v_x^2 + v_y^2) = 0 \tag{3.5.1}$$

We shall employ the complex notation $z = x + iy$, $\overline{z} = x - iy$, $u_z = 1/2(u_x - iu_y)$, etc.
From (3.5.1) and (3.4.3) we obtain

$$|u_{z\overline{z}}| \leq c|u_z| \leq K \tag{3.5.2}$$

since $u \in C^2(B)$.

If now $u_z(z_o) = 0$ for some $z_o \in B$, we can use the Lemma of Hartman - Wintner (cf. Lemma 3.6 below), to obtain the asymptotic representation

$$u_z = a(z - z_o)^n + o(|z - z_o|^n) \tag{3.5.3}$$

for some $a \in \mathbb{C}$, $a \neq 0$, and some positive integer n, unless $u_z \equiv 0$ in a neighborhood of z_o. The latter is not possible, however, since it implies that the set where $u_x v_y - u_y v_x = 0$ is nonvoid and open in B, and therefore $u_x v_y - u_y v_x \equiv 0$ in B in contradiction to the fact that τ is a surjective $C^{2,\alpha}$ map onto S. We can choose the local coordinates in such a way that

$$g_{ij}(\tau(z_o)) = \delta_{ij} \tag{3.5.4}$$

Using (3.4.3), (3.5.4) and integrating (3.5.3), we infer

$$w(z) := u + iv = \rho(z - z_o)^{n+1} + \sigma(\bar{z} - \bar{z}_o)^{n+1} + o(|z - z_o|^{n+1}) + w_o ,$$

where $\rho, \sigma \in \mathbb{R}$, $|\rho| + |\sigma| \neq 0$, $w_o = (u + iv)(z_o)$, in a neighborhood of z_o.
Without loss of generality, by performing homeomorphic linear transformations, we can assume $\rho = 1$, $\sigma > 0$, $z_o = w_o = 0$, i.e.

$$w(z) = z^{n+1} + \sigma\bar{z}^{n+1} + o(|z|^{n+1}) \tag{3.5.5}$$

If z_o is on the boundary ∂B, we can introduce local coordinates u,v on a neighborhood V of $\tau(z_o)$ with

$$v = 0 \quad \text{on } \partial S \cap V \tag{3.5.6}$$

and again

$$g_{ij}(\tau(z_o)) = \delta_{ij} \tag{3.5.4}$$

Performing an elementary Möbius transformation of the domain, we can assume that the boundary circle of B containing z_o is mapped onto the real axis (the x-axis in our previous notation) and B is contained in the upper half plane, i.e. $x > 0$. We now put

$$U(z) := \begin{cases} u(z) & \text{when } \operatorname{Re} z = x > 0 \\ -u(\bar{z}) & \text{when } \operatorname{Re} z < 0 \end{cases}$$

whenever defined, e.g. on a neighborhood $N(z_0)$.
Since (3.5.1) again holds for u , we conclude from (3.4.3)

$$|U_{z\bar{z}}| \le c|U_z| \le K \qquad (\text{since } u \in C^1(\bar{B}))$$

on $N(z_0) \setminus \{x = 0\}$.

Applying again Lemma 3.6 below, we again obtain the representation (3.5.3), i.e.

$$u_z = a(z - z_0)^n + o(|z - z_0|)^n$$

and

(3.5.7) $$w(z) = z^{n+1} + \sigma\bar{z}^{n+1} + o(|z|^{n+1})$$

after normalization in a neighborhood of $z_0 = 0$.
This , however, is in contradiction to the consequence we have obtained from the assumption that τ is not injective, namely that a whole continuum of points is mapped to a single point. This proves the lemma. (The application of the Hartman - Wintner formula in the above argument is due to E. Heinz [Hz7]).

Lemma 3.5: Under the assumptions of Lemma 3.4 , τ is a diffeomorphism between $\bar{B}$ and S .

Proof: We shall use an idea of Berg [Bg] to show that , if τ is a homeomorphism, a representation like (3.5.5) with $n \ge 1$ cannot hold and consequently the functional determinant cannot vanish. Thus the assertion follows from Lemma 3.4.
We shall present the argument only at boundary points, since the interior case is similar.
Thus, we assume for contradiction that (3.5.7) holds (with all the chosen normalizations, in particular $B \subset \{x > 0\}$).
We infer from (3.5.7)

$$u(re^{i\theta}) = (1 + \sigma)\, r^{n+1} \cos((n+1)\theta) + o(r^{n+1}),$$

and in particular

$$(3.5.8) \qquad u(re^{i\pi\, k/n+1}) = (1 + \sigma)\, r^{n+1}\, (-1)^k + o(r^{n+1})$$

for $k = 0,1,\dots n+1$.
For sufficiently small $\varepsilon > 0$ and $r \leqq \varepsilon$, the sign of the left hand side of (3.5.8) is therefore $(-1)^k$.
If z traverses a Jordan curve in the closed upper half plane $\{x \geqq 0\}$ which contains a piece of the real line $\{x = 0\}$ in a neighborhood of z_0 and is contained in $\{|z - z_0| \leqq \varepsilon\}$, then $u(z)$ changes sign at least $(n+1)$ times. Since τ is a homeomorphism by Lemma 3.4, we can choose ε and δ so small, that

$$(u + iv)^{-1}\{w \in (u + iv)\{x \geqq 0\},\ |w| \leqq \delta\} \leqq \{z = x + iy\colon x \geqq 0,\ |z| \leqq \varepsilon\}.$$

Therefore, the preimage γ of

$$\{w \in (u + iv)\{x \geqq 0\},\ |w| = \delta/2\}$$

is a Jordan curve with the properties stated above. However, as z traverses γ, $u(z)$ changes sign exactly once, contradicting $n \geqq 1$.
Thus $u_z \neq 0$ at z_0, and by (3.4.3) also $v_z \neq 0$. Since we already proved in Lemma 3.4, that τ is a homeomorphism, we conclude that τ is a diffeomorphism of $\overline{B}$ onto S.

q.e.d.

3.6. Proof of Theorem 3.1, continued

We are now in a position to finish the proof of Theorem 3.1, i.e. to prove that we can find a conformal homeomorphism τ between $\overline{B}$ and S, and that τ is a diffeomorphism in case $S \in C^{1,\alpha}$.
For this, we choose a sequence (g^n_{ij}) of metrics of class $C^{2,\alpha}$ on G, such that g^n_{ij} converges almost everywhere to g_{ij} as $n \to \infty$. By sections 3.3 and 3.4 and Lemma 3.5, we find a corresponding sequence of

diffeomorphisms $\tau_n : \overline{B_n} \to \overline{G}$, where B_n is again a circular domain as above and τ_n is conformal with respect to (g_{ij}^n). The maps τ_n satisfy uniform $H^{1,2}$ as well as C^α estimates by elliptic regularity theory,[1) and therefore we can find a subsequence, again denoted by (τ_n), with the property that the corresponding domains B_n converge to a circular domain B , and the τ_n converge uniformly and weakly in H_2^1 to a map τ , and τ is a weak solution of (3.4.3).
Furthermore, since the τ_n are diffeomorphisms, their inverses τ_n^{-1} satisfy a system of the same type as (3.4.3), namely

$$(3.6.1) \qquad y_u^n = \frac{g_{12}^n}{\sqrt{g^n}} x_u^n - \frac{g_{11}^n}{\sqrt{g^n}} x_v^n$$

$$y_v^n = \frac{g_{22}^n}{\sqrt{g^n}} x_u^n - \frac{g_{12}^n}{\sqrt{g^n}} x_v^n \quad ,$$

where $g^n = g_{11}^n g_{22}^n - (g_{12}^n)^2$.
Therefore, also τ_n^{-1} satisfies a uniform Hölder estimate by elliptic regularity theory, and thus we see that the limit map τ has to be invertible, i.e. a homeomorphism.
In case $S \in C^{1,\alpha}$, the metrics (g_{ij}^n) can be chosen to converge with respect to the C^α -norm to (g_{ij}). From (3.6.1) we infer that the τ_n^{-1} then satisfy uniform $C^{1,\alpha}$ estimates, and consequently the limit map τ is a diffeomorphism.
Thus we have found the desired conformal representation of S , and the proof of Theorem 3.1 is complete.

3.7. Uniqueness of conformal representations

Remark: Actually, it is not difficult to see, that each map τ constructed from the process of sections 3.3 - 3.4 has to be homeomorphism (and a diffeomorphism in the classical case $S \in C^{1,\alpha}$), since two such mappings τ_1 and τ_2 differ only by composition with an analytic function f , if one of them, say τ_1 , is a homeomorphism, i.e. $\tau_2(z) = f(\tau_1(z))$.
(This assertion is trivial in the classical case, and the general case follows by an approximation argument, cf. Lemma 8 , p. 274 in [BJS]). Then one can use the argument of Lemma 3.4 to show that also τ_2 is a homeomorphism, if it is the uniform limit of homeomorphisms.

1) These results are basically due to Morrey [M1].

Since a conformal automorphism of a subdomain B of the unit disc with three fix points on the boundary is the identity, we infer furthermore that the conformal representation τ constructed in 3.3 and 3.4 is unique if we require a three - point condition. Namely, for any two such representations τ_1 and τ_2 , $\tau_1^{-1} \circ \tau_2$ is a conformal automorphism of B with three fixed points on ∂B and thus has to be the identity map of B.

3.8. Applications of Theorem 3.1

In this section, we briefly mention some applications of Thm. 3.1. We call the case where S is at least of class $C^{1,\alpha}$ the classical case.

In [M3], §9.4, Morrey used the classical case of Thm. 3.1 to solve the k - contour Plateau problem. He showed, k Jordan curves in $\mathbb{R}^3$ (or a homogeneously regular Riemannian manifold) bound a k - connected minimal surface, if the minimum of area in the class of k - connected surfaces spanning the boundary configuration is less than in the class of unions of surfaces of lower connectivity. Thus, since his proof of Thm. 3.1 was not correct, we have completed his approach to the k - contour Plateau problem.
The nonclassical 1 - contour case of Thm. 3.1 was used in the approach of Ahlfors and Bers to Teichmüller theory, cf. [A1], [A2], [AB], and [B1]. An essential notion for this approach is the concept of a Beltrami differential, i.e. $\mu \in L^\infty(D,\mathbb{C})$ with $|\mu|_\infty < 1$. μ defines a conformal structure on D , represented by the metric

$$|dz + \mu d\bar{z}|^2 ,$$

and a conformal map from the usual structure on D onto the one given by μ is a solution of the Beltrami equation

$$(3.8.1) \qquad w_{\bar{z}} = \mu w_z$$

Given a compact Riemann surface Σ of genus at least 2 , the universal cover of Σ is conformally the unit disc, and Σ is conformally equivalent to a fundamental domain in D , and $\pi_1(\Sigma)$ gives a group Γ of covering transformations which are conformal automorphisms of D.

The mentioned approach to Teichmüller theory then proceeds by showing existence of solutions of (3.8.1) for given μ (which follows from Thm. 3.1) and examining its dependance on variations of μ (a modern approach to this could use a - priori estimates and uniqueness of solutions of (3.7.1), cf. 3.1, 3.4, 3.7). Finally it is shown that the Teichmüller space of Σ , i.e. the space of moduli of conformal structures on Σ together with a given marking, is the quotient of the space of those Beltrami differentials on D which are invariant under Γ , i.e. those which represent conformal structures on Σ , divided by the subspace of those which are orthogonal with respect to a suitable inner product to the space of holomorphic quadratic differentials on Σ .

Finally, the approach of Earle - Eells [EE] to Teichmüller theory needs only the classical one - contour case of Thm. 3.1 together with a theorem about the continuous dependance of solutions of (3.8.1) on variations of μ (cf. [ESz]), which can be proved via uniqueness and a - priori estimates. They employ the point of view of Kodaira - Spencer about complex structures and thus have to deal only with smooth μ . We shall come back to Teichmüller theory in 12.3 under different aspects.

Our main application of Thm. 3.1 in the present book, however, will consist in deriving existence and regularity of harmonic diffeomorphisms, cf. in particular chapters 6, 7 and 8 .

3.9. The Hartman - Wintner Lemma

In this section, we shall prove the Lemma of Hartman - Wintner ([HtW], pp. 455 - 458) in the form which was used in the preceding paragraphs. We shall use the complex notation, i.e.

$$z = x + iy , \quad u_z = \frac{1}{2}\left(\frac{\partial}{\partial x} u + i \frac{\partial}{\partial y} u\right), \quad u_{\bar z} = \frac{1}{2}\left(\frac{\partial}{\partial x} u - i \frac{\partial}{\partial y} u\right), \text{ etc.}$$

Lemma 3.6: Suppose $u \in C^{1,1}(D,\mathbb{R})$ satisfies almost everywhere

$$(3.9.1) \qquad |u_{z\bar z}| \le K(|u_z| + |u|)$$

where K is a fixed positive constant.

If

$$(3.9.2) \qquad u(z) = o(|z|^n) \quad \text{for some } n \in \mathbb{N}, \text{ then}$$

$$(3.9.3) \qquad \lim_{z \to 0} u_z \cdot z^{-n} \quad \text{exists.}$$

If (3.9.2) holds for all $n \in \mathbb{N}$, then

$$u \equiv 0 .$$

Proof: First we note that since $u \in C^{1,1}$, $u_{z\bar{z}}$ exists almost everywhere by Rademacher's Theorem. If B is a closed subdomain of D with Lipschitz boundary ∂B and $g \in C^1(B,\mathbb{C})$, then

$$(3.9.4)\qquad \oint_{\partial B} g u_z = \int_B (u_z g_{\bar{z}} + u_{z\bar{z}} g)\,dz$$

Suppose that

$$(3.9.5)\qquad u_z = o(|z|^{k-1}) \text{ for some } k \in \mathbb{N}$$

In (3.9.4), we take $B = \{z: \varepsilon \le |z| \le R\ ,\ |z-z_o| \ge \varepsilon\}$, $R < \min(1,\frac{\pi}{4K})$, and $g = z^{-k}(z-z_o)^{-1}$, $z_o \neq 0$. Since $g_{\bar{z}} = 0$ in B , we obtain by letting $\varepsilon \to 0$ from (3.9.4)

$$(3.9.6)\qquad 2\pi\, u_z(z_o)\, z_o^{-k} = \oint_{|z|=R} u_z z^{-k}(z-z_o)^{-1} - \int_{B(0,R)} u_{z\bar{z}} z^{-k}(z-z_o)^{-1}\,dz$$

and hence from (3.9.1)

$$(3.9.7)\qquad 2\pi |u_z(z_o) z_o^{-k}| \le \oint_{|z|=R} |u_z z^{-k}(z-z_o)^{-1}|\,|dz| + K \int_{B(0,R)} (|u_z| + |u|)|z|^{-k}|z-z_o|^{-1}\,dz$$

We now want to estimate

$$I_k := \int_{B(0,R)} |u_z||z|^{-k}|z-z_o|^{-1}\,dz$$

uniformly as $z_o \to 0$. If that can be achieved, then the second integral in (3.9.6) converges to some limit as $z_o \to 0$, and hence

$$\lim_{z\to 0} u_z \cdot z^{-k} \qquad \text{exists.}$$

Moreover, if $k < n$, then this limit vanishes because of (3.9.2), and hence (3.9.5) holds for $k+1$. On the other hand, the first assertion of the lemma is trivial, if $n = 0$, and $n \ge 1$ implies (3.9.5) for $k = 1$. Thus, by induction, (3.9.5) holds for $k = n$, which implies the first assertion of the lemma.

In order to control I_k , we multiply (3.9.7) by $|z_o - z_1|^{-1}$, use

(3.9.8) $$\int_{B(0,R)} |z - z_0|^{-1}\, dz < 2R , \quad \text{if } |z_0| < R \text{ and}$$

$$(z - z_0)^{-1}(z_0 - z_1)^{-1} = (z - z_1)^{-1}((z - z_0)^{-1} + (z_0 - z_1)^{-1}) ,$$

and obtain, integrating w.r.t. z_0 , $|z_0| < R$,

$$2\pi \int_{B(0,R)} |u_z| \cdot |z^{-k}||z - z_0|^{-1}\, dz \le$$

(3.9.9) $$\le 4R \oint_{|z|=R} |u_z z^{-k} (z - z_1)^{-1}||dz| +$$

$$+ 4KR \int_{B(0,R)} (|u_z| + |u|)|z|^{-k}|z - z_1|^{-1}\, dz$$

By choice of $R(R < \frac{\pi}{2K})$, and since $k < n$, i.e. $|uz^{-k}| = o(|z|)$ by assumption, (3.9.9) controls I_k , putting $z_1 = z_0$.
In particular, by induction, (3.9.9) holds for $k = n - 1$, and letting $z_1 \to 0$, we obtain

(3.9.10)
$$(2\pi - 4KR) \int_{B(0,R)} |u_z z^{-n}|dz \le$$

$$\le 4R \oint_{|z|=R} |u_z z^{-n}||dz| + 4KR \int_{B(0,R)} |u||z|^{-n}\, dz$$

Now $|u(z)| \le \int_0^1 |zu_z(tz)|dt$, which leads to

$$\int_{B(0,R)} |uz^{-n}|dz \le \int_0^1 \int_{B(0,R)} |z^{-n+1} u_z(tz)|dzdt =$$

$$= \int_0^1 t^{n-3} \int_{B(0,tR)} |z^{-n+1} u_z(z)|dzdt$$

Hence for $n \ge 3$

$$\int_{B(0,R)} |uz^{-n}|dz \le \int_{B(0,R)} |u_z z^{-n+1}|\, dz .$$

Since $|z| \le R < 1$ by choice of R , we see that the second integral in (3.9.10) can be absorbed into the left hand side, yielding

$$(3.9.11)\qquad (2\pi - 8KR) \int_{B(0,R)} |u_z z^{-n}| dz \le$$

$$\le 2R \int_{|z|=R} |u_z z^{-n}| |dz|$$

Since $(2\pi - 8KR) > 0$ by choice of R, it is not difficult to see that (3.9.11) for all $n \in \mathbb{N}$ implies $u_z \equiv 0$ and hence $u \equiv 0$ which finishes the proof.
Namely, otherwise, there would exist z_o with $|z_o| < R$ and $u_z(z_o) = c \neq 0$, and the left hand side of (3.9.11) would grow like $c|z_o|^{-n}$ and the right hand side like $c_o R^{-n}$, c, c_o being independant of n, and hence (3.9.11) could not hold for all n.

q.e.d.

4. Existence theorems for harmonic maps between surfaces

4.1. A maximum principle for energy minimizing maps

We assume that Σ_1 and Σ_2 are twodimensional Riemannian manifolds. Ω is an open subset of Σ_1. $i(\Sigma)$ denotes the injectivity radius of Σ. In this chapter, we shall solve existence problems by an energy minimizing procedure. A useful tool will be the following maximum principle for energy minimizing maps which is taken from [J6] and based on the same idea as the one in [H1], Lemma 6.

Lemma 4.1: Suppose that B_o and B_1, $B_o \subset B_1$, are closed subsets of a Riemannian manifold N. Suppose that there exists a projection map

$$\pi: B_1 \to B_o ,$$

which is the identity on B_o and which is distance decreasing outside B_o, i.e.

$$d(\pi(x),\pi(y)) < d(x,y)$$

whenever $x,y \in B_1 \setminus B_0$, $x \neq y$.
If $h: \Omega \to B_1$ is an energy minimizing W_2^1 mapping with respect to fixed boundary values which are contained in B_0 , i.e.

(4.1.1) $\qquad h(\partial\Omega) \subset B_0$,

then we also have

$$h(\Omega) \subset B_0 ,$$

if we choose a suitable representant of the Sobolev mapping h .

Proof: Since π is Lipschitz continuous, it is easily seen that $|d\pi(v)| < |v|$ for almost every nonzero $v \in T_xN$, $x \in B_1 \setminus B_0$, and that $\pi \circ h \in W_2^1(\Omega,N)$.
Thus we would have

$$E(\pi \circ h) < E(h) ,$$

contradicting the minimality of h , unless $dh = 0$ a.e. on $h^{-1}(B_1 \setminus B_0)$. Thus $dh = d\pi \circ h$ a.e. on Ω , and since h and $\pi \circ h$ agree on $\partial\Omega$ by (4.1.1) , we conclude from the Poincaré inequality that $\pi \circ h = h$ a.e. on Ω , which easily implies the claim.

Lemma 4.2: Suppose that B_0 and B_1 , $B_0 \subset B_1$, are compact subsets of a Riemannian manifold N , and that every point in $B_1 \setminus B_0$ can be joined to ∂B_0 by a unique geodesic normal to ∂B_0 , and that the distance between every pair of such geodesics normal to ∂B_0 is in $B_1 \setminus B_0$ always bigger than on ∂B_0 . Then the same conclusion as in Lemma 4.1 holds.

Proof: We project $B_1 \setminus B_0$ along normal geodesics onto ∂B_0 and apply Lemma 4.1.

q.e.d.

Another consequence of Lemma 4.1 is

Lemma 4.3: Suppose B_0 is a geodesic ball with radius s and center p.

$s \leq 1/3 \min (i(p), \pi/2\kappa)$, where κ^2 is an upper bound for the sectional curvature of N and i(p) is the injectivity radius of p .
If h: $\Omega \to N$ is energy minimzing among maps which are homotopic to some map g: $\Omega \to B_o$, and if $h(\partial\Omega) \subset B_o$, then also

$$h(\Omega) \subset B_o .$$

(for a suitable representative of h , again).

Proof: By assumption, we can introduce geodesic polar coordinates (r,φ) on $B(p,3s)$ $(0 \leq r \leq 3s)$.
We define a map π in the following way:

$$\begin{aligned}
\pi(r,\varphi) &= (r,\varphi) && \text{if } r \leq s \\
\pi(r,\varphi) &= (\tfrac{1}{2}(3s-r),\varphi) && \text{if } s \leq r \leq 3s \\
\pi(q) &= p && \text{if } q \in N \setminus B(p,3s)
\end{aligned}$$

(Here, we have identified a point in B(p,3s) with its representation in geodesic polar coordinates.)
Using the Rauch comparison theorem, it is easily seen, that π satisfies the assumptions of Lemma 4.1.

q.e.d.

4.2. The Dirichlet problem, if the image is contained in a convex disc

We now want to prove the following result of Hildebrandt - Kaul - Widman [HKW3], using the Courant - Lebesgue - Lemma 3.1 and Lemma 4.2. (Actually, most of it already follows from Morrey's work [M2].)

Theorem 4.1: Suppose $\partial\Omega \neq \emptyset$, B(p,M) is a disc in Σ_2 with radius $M < \frac{\pi}{2\kappa}$, where $\kappa^2 \geq 0$ is an upper bound of the Gauss curvature of B(p,M), and $g: \partial\Omega \to B(p,M)$ is continuous and admits an extension $\bar{g} \in H_2^1(\Omega, B(p,M))$.[1)]

Then there exists a harmonic map h: $\Omega \to B(p,M)$ with boundary values g and h minimizes the energy with respect to these boundary values. Vice versa, each such energy minimizing map is harmonic. The modulus of continuity of h can be estimated in terms of λ , $i(\Sigma_1)$, M , κ , and $E(\bar{g})$ and the modulus of continuity of g .

---

1) Here, we can define $H_2^1(\Omega, B(p,M))$ unambiguously with the help of the global coordinates on B(p,M) given by $\exp_p$.

Proof: (the idea is taken from the proof of Thm. 4.1 in [HW 1].)
Since the cut locus of a point p is a closed set, we can find some M^1, $M < M^1 < \pi/2\kappa$ for which $B(p,M^1)$ is still a disc (cf. Thm. 2.1). We take a minimizing sequence for the energy in $V: = \{v \in H_2^1(\Omega, B(p,M^1)),\ v_{|\partial\Omega} = g\}$. Such a sequence has a subsequence converging weakly in H_2^1, and the limit, denoted by h, minimizes energy in its class because of the lower semicontinuity of the Dirichlet integral. Applying Lemma 4.2 to $B_0 = B(p,M)$, $B_1 = B(p,M^1)$, we conclude that h actually maps Ω into the smaller ball $B(p,M)$.
By Thm. 2.1, every two points in $B(p,M)$ can be joined by a unique geodesic arc in $B(p,M)$, and this arc is free of conjugate points. Therefore, we can apply the Rauch comparison theorem in the following way. Suppose that $q \in B(p,M)$, v_1 and v_2 are unit vectors in $T_q\Sigma_2$, and c_1, c_2 are the geodesics parametrized by arclength and starting at q with tangent vectors v_1, v_2 resp. Then

$$|v_1 - v_2|\kappa^{-1} \sin(t\kappa) \leq d(c_1(t), c_2(t))$$

as long as $c_1(t)$, $c_2(t) \in B(p,M)$.
Therefore, on $B(p,M) \setminus B(q,\varepsilon)$,

$$d(c_1(t),c_2(t)) \geq \min\left(d(c_1(\varepsilon),c_2(\varepsilon)),\ |v_1-v_2|\kappa^{-1}\sin(2M\kappa)\right),$$

Consequently, there exists $\varepsilon_0 > 0$ with the property that $B_0: = B(q,\varepsilon) \cap B(p,M)$ and $B_1: = B(p,M)$ satisfy the assumptions of Lemma 4.2 for every $q \in B(p,M)$ and every $\varepsilon \leq \varepsilon_0$. Lemma 3.1 then implies that for each $x \in \Omega$ there exists a sufficiently small $\rho > 0$ with the property that

$$h(B(x,\rho) \cap \Omega) \subset B(q,\varepsilon)$$

for some $q \in B(p,M)$. ρ depends on ε, λ, $i(\Sigma_1)$, the energy of h (which is bounded by the energy of $\bar{g}$), and the modulus of continuity of g.
Therefore, Lemma 4.2 implies the continuity of h. Furthermore, it is easy to see (cf. [HKW3]) that h is a weak solution of $\tau(h) = 0$, since h is energy minimizing in V and on the other hand, $h(\Omega) \subset B(p,M)$, so that h actually lies in the interior of V, and the variation of E at h in the direction of any $\psi \in \mathring{H}_2^1 \cap L^\infty(\Omega,\Sigma_2)$ has to vanish. Higher regularity follows e.g. from [LU] and classical linear elliptic theory (cf. also [G] and [Hi2]). (We shall derive higher order

estimates in the following chapters in much more detail, however).

q.e.d.

4.3. Remarks about the higher-dimensional situation

Thm. 4.1 holds in arbitrary dimensions, again due to Hildebrandt-Kaul-Widman [HKW3]. A different proof, using the heat flow method, was given in [J4]. While conceptually more complicated than the first one (which proves the regularity of the minimum of energy) it leads to the following maximum principle (cf. [J5]).

Proposition 4.1: Suppose $f: \Omega \subset N$ is a harmonic mapping from a bounded domain Ω in some Riemannian manifold into a complete Riemannian manifold. Assume $f(\Omega) \subset B(p,M)$, where again $M < \pi/2\kappa$ and $B(p,M)$ is disjoint to the cut locus of p.
If $f(\partial\Omega) \subset Y$, where $Y \subset B(p,M)$ has a smooth convex boundary, then also

$$f(\Omega) \subset Y .$$

One can also derive a-priori estimates, e.g. for the modulus of continuity, for harmonic maps with image contained in a strictly convex ball in arbitrary dimensions, cf. e.g. [HJW] or [GH]. The proof, however, is much more involved.[1] The idea can be roughly outlined as follows: First of all, the estimates are somewhat easier, if the stronger condition $M < \pi/4\kappa$ is supposed, because then the squared distance function from any $q \in B(p,M)$ is strictly convex on $B(p,M)$. The general case $M < \pi/2\kappa$ then has to be covered by an iteration argument using first that for q close to p $d^2(\cdot,q)$ is still strictly convex on $B(p,M)$. From this one can gain enough informations to conclude that, if one shrinks the domain somewhat, for some more q $d^2(\cdot,q)$ is strictly convex on $B(p,M)$ and so on. The idea to iterate is due to Wiegner [Wi]. We shall exhibit a version of such an argument at boundary points, where it is somewhat easier, in section 6.5.
A somewhat more geometric version of the proof is given in [GJ]. A different approach is due to Sperner [Sp]. He uses the fact that the functions constructed by Jäger and Kaul [JäK2] for the proof of their uniqueness theorem (cf. 5.2) are strictly convex on the whole of $B(p,M)$ and can have a minimum at an arbitrarily assigned point. Therefore, the argument for $M < \pi/4\kappa$ can cover the more general case if one uses these functions instead of the squared distance. Still other proofs

[1] The method of proof is due to Hildebrandt-Widman [HW2].

were provided by Eliasson [Es] and Tolksdorf [To] .
For a general account of the regularity results for the class of quasilinear elliptic systems which includes that one for harmonic maps, we refer to the lecture notes [Hi2] by Hildebrandt.
More general existence results can be obtained by the approach of partial regularity (cf. [SU1], [SU2],[GG1],[GG2],[J6],[JM], [E]). This approach first characterizes the possible singularities of energy minimizing maps and then shows that under suitable geometric hypotheses these singularities cannot occur. The best result can be obtained with the methods of Schoen - Uhlenbeck. With their techniques, one can solve the Dirichlet problem, if the image has a strictly convex boundary and supports a strictly convex function. Using the methods of Giaquinta - Giusti, one can prove the same result, provided the image is covered by a single coordinate chart (cf. [J6]). This last condition is not as restrictive as one might first think, since a manifold with strictly convex boundary and supporting a strictly convex function is contractible anyway. For, if it would have some nonvanishing homotopy group, a well known argument from the theory of closed geodesics (cf. [O]) and the strict convexity of the boundary would yield a nontrivial closed geodesic inside, in contradiction to the existence of the strictly convex function, cf. Prop. 5.1.
The Dirichlet problem for the case, where the image has nonpositive sectional curvature was previously solved by Hamilton, using the heat flow method ([Hm]).
In two dimensions, however, one does not need any geometric restrictions on the image like convexity conditions to prove the existence of harmonic mappings. The only condition needed is of a topological nature, namely that the second homotopy group of the image has to vanish, as we shall see in the following sections.

4.4. The Theorem of Lemaire and Sacks - Uhlenbeck

We are now in a position to attack the general existence problem for harmonic maps between surfaces.
For this purpose, let Σ_1 and Σ_2 denote compact surfaces, $\partial\Sigma_2 = \emptyset$, but Σ_1 possibly having nonempty boundary. Let $\varphi: \Sigma_1 \to \Sigma_2$ be a continuous map with finite energy. We denote by $[\varphi]$ the class of all continuous maps which are homotopic to φ and coincide with φ on $\partial\Sigma_1$, in case $\partial\Sigma_1 \neq \emptyset$.
We choose $s = 1/3 \min (i(\Sigma_2)\ ,\ \pi/2\kappa)$, where $\kappa^2 \geq 0$ is an upper curvature bound on Σ_2 , and $i(\Sigma_2)$ is the injectivity radius of Σ_2 .

Let $\delta_0 \leq \min(1, i(\Sigma_1)^2, 1/\lambda^2)$ ($-\lambda^2$ being a lower bound for the curvature of Σ_1) satisfy

$$(4.4.1) \qquad 2\pi \cdot E(\varphi)^{1/2} (\log 1/\delta_0)^{-1/2} \leq s/2 ,$$

where $E(\varphi)$ is the energy of φ , and

$$(4.4.2) \qquad d(x_1,x_2) \leq \sqrt{\delta_0} \Rightarrow d(\varphi(x_1),\varphi(x_2)) \leq s/2 \quad \text{for } x_1,x_2 \in \partial\Sigma_1 .$$

Let $0 < \delta < \delta_0$. There exists a finite number of points $x_i \in \Sigma_1$, $i = 1,\dots m = m(\delta)$, for which the discs $B(x_i,\delta/2)$ cover Σ_1 .
We let u_n be a continuous energy minimizing sequence in $[\varphi]$, $E(u_n) \leq E(\varphi)$ w.l..o.g. for all n .
Applying Lemma 3.1 and using (4.4.1) and (4.4.2), for every n , we can find $r_{n,1}$, $\delta < r_{n,1} < \sqrt{\delta}$, and $p_{n,1} \in \Sigma_2$ with the property that

$$(4.4.3) \qquad u_n(\partial B(x_1,r_{n,1})) \subset B(p_{n,1},s)$$

(Here, we have defined $B(x,r) = \{y \in \Sigma_1 : d(x,y) \leq r\}$ and thus $\partial B(x,r) = \partial(B(x,r) \cap \Sigma_1)$ near the boundary.)
We now have two possibilities:
either

1) There exists some δ , $0 < \delta \leq \delta_0$, with the property that for any $x \in \Sigma_1$, some r , depending on x and n with $\delta < r \leq \sqrt{\delta}$ and with $u_n(\partial B(x,r)) \subset B(p,s)$ for some $p \in \Sigma_2$, and every sufficiently large n $u_n|B(x,r)$ is homotopic to the solution of the Dirichlet problem

$$(4.4.4) \qquad \begin{aligned} &g: B(x,r) \to B(p,s) \qquad \text{harmonic and energy minimizing} \\ &g|\partial B(x,r) = u_n|\partial B(x,r) \end{aligned}$$

   (The existence of g is ensured by Thm. 4.1; g is actually unique by Thm. 5.1, but this is not needed in the following constructions)

or

2) Possibly choosing a subsequence of the u_n , we can find a sequence of points $x_n \in \Sigma_1$, and radii $r_n > 0$, $x_n \to x_0 \in \Sigma_1$, $r_n \to 0$, with $u_n(\partial B(x_n,r_n)) \subset B(p_n,\varepsilon_n)$ for some $p_n \in \Sigma_2$, $p_n \to p \in \Sigma_2$, $\varepsilon_n \to 0$ (using Lemma 3.1), but for which $u_n|B(x_n,r_n)$ is not homotopic to the solution of the Dirichlet problem (4.4.4).

In case 1), we replace u_n on $B(x_1,r_{n,1})$ by the solution of the Dirichlet problem (4.4.4) for $x = x_1$ and $r = r_{n,1}$. We can assume $r_{n,1} \to r_1$ and, using the interior modulus of continuity estimates for the solution of (4.4.4) (cf. Thm. 4.1) that the replaced maps, denoted by u_n^1, converge uniformly on $B(x_1,\delta-\eta)$, for any $0 < \eta < \delta$. By Lemma 4.3

$$(4.4.5) \qquad E(u_n^1) \le E(u_n) .$$

By the same argument as above, we then find radii $r_{n,2}$, $\delta < r_{n,2} < \sqrt{\delta}$, with

$$u_n^1(\partial B(x_2,r_{n,2})) \subset B(p_{n,2},s)$$

for points $p_{n,2} \in \Sigma_2$.

Again, we replace u_n^1 on $B(x_2,r_{n,2})$ by the solution of the Dirichlet problem (4.4.4) for $x = x_2$ and $r = r_{n,2}$. We denote the new maps by u_n^2. Again, w.l.o.g., $r_{n,2} \to r_2$.

If we take into consideration that, by the first replacement step, u_n^1 in particular converges uniformly on $B(x_2,r_2) \cap B(x_1,\delta-\eta/2)$, if $0 < \eta < \delta$, we see that the boundary values for our second replacement step converge uniformly on $\partial B(x_2,r_{n,2}) \cap B(x_1,\delta-\eta/2)$.

Using the estimates for the modulus of continuity for the solution of (4.4.4) at these boundary points (cf. Thm. 4.1), we can assume that the maps u_n^2 converge uniformly on $B(x_1,\delta-\eta) \cup B(x_2,\delta-\eta)$, if $0 < \eta < \delta$.

Furthermore, by Lemma 4.3 again and (4.4.5)

$$E(u_n^2) \le E(u_n^1) \le E(u_n) .$$

In this way, we we repeat the replacement argument, until we get a sequence $u_n^m =: v_n$, with

$$(4.4.6) \qquad E(v_n) \le E(u_n)$$

which converges uniformly on all balls $B(x_i,\delta/2)$, $i = 1,...m$, and hence on all of Σ_1, since these balls cover Σ_1.

We denote the limit of the v_n by u. By uniform convergence, u is homotopic to φ.

Since $E(v_n) \le E(\varphi)$ by (4.4.6), we can assume that the v_n converge also weakly in H^1_2 to u, and by lower semicontinuity of the energy w.r.t. weak H^1_2 convergence and since the v_n are a minimizing sequence by (4.4.6), u minimizes energy in its homotopy class.

In particular, u minimizes energy when restricted to small balls, and hence it is harmonic and regular by Lemma 4.3 and Thm. 4.1.

Observing that if $\pi_2(\Sigma_2) = 0$, any two maps from a disc into Σ_2 are homotopic, we obtain

Theorem 4.2: Suppose Σ_1 and Σ_2 are compact surfaces, $\partial\Sigma_2 = \emptyset$, and $\pi_2(\Sigma_2) = 0$. If $\varphi : \Sigma_1 \to \Sigma_2$ is a continuous map with finite energy, then there exists a harmonic map $u: \Sigma_1 \to \Sigma_2$ which is homotopic to φ, coincides with φ on $\partial\Sigma_1$ in case $\partial\Sigma_1 \neq \emptyset$ and is energy minimizing among all such maps.

Theorem 4.2 is the fundamental existence theorem due to Lemaire ([L1], [L2]) and Sacks - Uhlenbeck ([SkU], in case $\partial\Sigma_1 = \emptyset$).

A different proof was given by Schoen - Yau [SY2].

The present proof is taken from [J6].

In the case of the Dirichlet problem, it is actually not necessary that Σ_2 is compact, but only that it is homogeneously regular in the sense of Morrey [M2], cf. [L2]. Looking at the present proof, we only have to observe that the fixed boundary values imply the uniform boundedness of the equicontinuous sequence v_n, and hence we can select a uniformly convergent subsequence as in the case of a compact image.

4.5. The Dirichlet problem, if the image is homeomorphic to S^2. Solution, if the boundary values are nonconstant.

In this section, we want to show the following result of Jost [J7] and Brezis and Coron [BC2] (in the latter paper, only simply connected domains are treated).

Theorem 4.3: Suppose Σ_1 is a compact two - dimensional Riemannian

<u>manifold with nonempty boundary</u> $\partial\Sigma_1$, <u>and</u> Σ_2 <u>is a Riemannian manifold homeomorphic to</u> S^2 <u>(the standard 2-sphere), and</u> $\psi: \partial\Sigma_1 \to \Sigma_2$ <u>is a continuous map, not mapping</u> $\partial\Sigma_1$ <u>onto a single point and admitting a continuous extension to a map from</u> Σ_1 <u>to</u> Σ_2 <u>with finite energy.</u>
<u>Then there are at least two homotopically different harmonic maps</u> $u: \Sigma_1 \to \Sigma_2$ <u>with</u> $u|\partial\Sigma_1 = \psi$, <u>and both mappings minimize energy in their respective homotopy classes.</u>

<u>Proof</u>: We first investigate more closely case 2). W.l.o.g. $B(p_n,\varepsilon_n) \subset B(p,2\varepsilon_n)$ and $\varepsilon_n \leq s/2$ for all n , and thus the solution g of (4.4.4) for $x = x_n$, $r = r_n$ is contained in $B(p,2\varepsilon_n)$ by Lemma 4.3.
Since $u_n|B(x_n,r_n)$ is not homotopic to g , it has to cover $\Sigma_2 \setminus B(p,2\varepsilon_n)$.
If we define

$$\tilde{u}_n = \begin{cases} u_n & \text{on } \Sigma_1 \setminus B(x_n,r_n) \\ g & \text{on } B(x_n,r_n) \end{cases},$$

then we see that

$$\text{(4.5.1)} \qquad \lim E(u_n) \geq \lim E(u_n|\Sigma_1 \setminus B(x_n,r_n)) + \lim E(u_n|B(x_n,r_n))$$

$$\geq \lim E(\tilde{u}_n) + \text{Area } (\Sigma_2) ,$$

since $E(g) \to 0$ as $n \to \infty$, because

$$\int_0^{2\pi} |g_\theta(r_n,\theta)|^2 d\theta \to 0$$

as $n \to \infty$. (cf. (3.2.3)).
(Furthermore, it is elementary that

$$E(v|B) \geq \text{Area } (v(B)),$$

and equality holds if and only if v is conformal).
We now define

$$E_\alpha : = \inf \{E(v): \quad v \in \alpha\}$$

for a homotopy class α of maps with $v|\partial\Sigma_1 = \psi$,
and

$$E: = \min_\alpha E_\alpha .$$

We first show the existence of a minimizing harmonic map in any homotopy class α with

$$(4.5.2) \qquad E_\alpha < E + \text{Area}\ (\Sigma_2)$$

We choose a minimizing sequence u_n in α with

$$E(u_n) < E + \text{Area}\ (\Sigma_2).$$

Assuming that 2) holds, we define $\tilde{u}_n$ as above. Since clearly

$$E(\tilde{u}_n) \geq E\ ,$$

this would contradict (4.5.1), however. Therefore, as shown above, we obtain an energy minimzing harmonic map in α (cf. [BC1] for a similar argument). Now let $\tilde{\alpha}$ be a homotopy class with

$$E_{\tilde{\alpha}} = E\ ,$$

and let $\tilde{u}$ an energy minimizing map in $\tilde{\alpha}$, i.e. $E(\tilde{u}) = E$.
We want to construct a map v in some homotopy class $\alpha \neq \tilde{\alpha}$ with

$$(4.5.3.) \qquad E(v) < E(\tilde{u}) + \text{Area}\ (\Sigma_2)\ .$$

Then the arguments above show that we can find a harmonic map of minimal energy in α . In order to complete the proof, it thus only remains to construct v .
The metric on Σ_2 is conformally equivalent to the standard metric on S^2 , and thus, we can use S^2 as a parameter domain for the image. Since ψ is not a constant map, also $\tilde{u}$ is not a constant map, and hence we can find a point x_o in the interior of Σ_1 for which $d\tilde{u}(x_o) \neq 0$. Rotating S^2 , we can assume that $\tilde{u}(x_o)$ is the south pole p_o . We introduce local coordinates on the image by stereographic projection $\pi\colon S^2 \to \mathbb{C}$ from the south pole p_o . $d\pi(p_o)$ then is the identity map up to a conformal factor. By Taylor's theorem, $\pi \circ \tilde{u}|\partial B(x_o,\varepsilon)$ is a linear map up to an error of order $O(\varepsilon^2)$, i.e.

$$(4.5.4) \qquad |\pi \circ \tilde{u}(x) - d(\pi \circ \tilde{u})(x_o)(x - x_o)| = O(\varepsilon^2)$$

for $x \in \partial B(x_o,\varepsilon)$.
We now look at conformal maps of the form

$$w = az + b/z \ , \quad a,b \in \mathbb{C} \ , \quad a = a_1 + ia_2 \ , \quad b = b_1 + ib_2 \ .$$

The restrictions of such a map to a circle $\rho(\cos\theta + i\sin\theta)$ in $\mathbb{C}$ is given by

$$u = (a_1\rho + \frac{b_1}{\rho})\cos\theta + (\frac{b_2}{\rho} - a_2\rho)\sin\theta$$

$$v = (a_2\rho + \frac{b_2}{\rho})\cos\theta + (a_1\rho - \frac{b_1}{\rho})\sin\theta \ ,$$

where $w = u + iv$.

Therefore, we can choose a and b in such a way, that w restricted to this circle coincides with any prescribed nontrivial linear map. This map is nonsingular if

$$\rho^4 \neq \frac{b_1^2 + b_2^2}{a_1^2 + a_2^2}$$

W.l.o.g.

$$\rho^4 \leq \frac{b_1^2 + b_2^2}{a_1^2 + a_2^2} \tag{4.5.5}$$

(otherwise we perform an inversion at the unit circle).

Hence w can be extended as a conformal map from the interior of the circle $\rho(\cos\theta + i\sin\theta)$ onto the exterior of its image. (If equality holds in (4.5.5), then this image is a straight line covered twice, and the exterior is the completement of this line in the complex plane).

We are now in a position to define v .

On $\Sigma_1 \setminus B(x_0,\varepsilon)$ we put $v = u$.

On $B(x_0,\varepsilon-\varepsilon^2)$ we choose a conformal w as above which coincides on the boundary with the linear map $\frac{1}{1-\varepsilon}\cdot d(\pi\circ\tilde{u})(x_0)$, and put $v = \pi^{-1}\circ w$.

On $B(x_0,\varepsilon) \setminus B(x_0,\varepsilon-\varepsilon^2)$ we interpolate in the following way. We introduce polar coordinates r,φ and define

$$f(\varphi) := (\pi\circ\tilde{u})(\varepsilon,\varphi)$$

$$g(\varphi) := d(\pi\circ\tilde{u})(\varepsilon,\varphi) = \frac{1}{1-\varepsilon}\, d(\pi\circ\tilde{u})(\varepsilon-\varepsilon^2,\varphi)$$

and

$$t(r,\varphi) := (f(\varphi) - g(\varphi))\cdot\frac{r}{\varepsilon^2} + \frac{1}{\varepsilon}(g(\varphi) - (1-\varepsilon)f(\varphi))$$

Thus $t(r,\varphi)$ coincides with $f(\varphi)$ and $g(\varphi)$, resp. for $r = \varepsilon$ and $r = \varepsilon - \varepsilon^2$, resp.

The energy of $t(r,\varphi)$ on the annulus $B(x_0,\varepsilon) \setminus B(x_0,\varepsilon-\varepsilon^2)$ is given by

$$E(t) = \int_{r=\varepsilon-\varepsilon^2}^{\varepsilon} \int_{\varphi=0}^{2\pi} \left(\frac{1}{\varepsilon^4}|f(\varphi) - g(\varphi)|^2 + \frac{1}{r^2}\left|\left(\frac{r}{\varepsilon^2} - \frac{1-\varepsilon}{\varepsilon}\right)f'(\varphi) + \right.\right.$$

$$\left.\left. + \left(\frac{1}{\varepsilon} - \frac{r}{\varepsilon^2}\right)g'(\varphi)\right|^2\right) r\,dr\,d\varphi$$

Using (4.5.4) and $|f'(\varphi)| = O(\varepsilon)$, $|g'(\varphi)| = O(\varepsilon)$, we calculate

$$E(t) = O(\varepsilon^3) \quad ,$$

and hence also,

$$E(\pi^{-1} \circ t) = O(\varepsilon^3).$$

We put $v = \pi^{-1} \circ t$ on the annulus $B(x_0,\varepsilon) \setminus B(x_0,\varepsilon-\varepsilon^2)$.
Therefore

$$E(v) = E(\tilde{u}|\Sigma_1 \setminus B(x_0,\varepsilon)) + E(\pi^{-1} \circ w|B(x_0,\varepsilon-\varepsilon^2)) +$$

$$+ E(\pi^{-1} \circ t|B(x_0,\varepsilon) \setminus B(x_0,\varepsilon-\varepsilon^2)) \leq$$

$$\leq E(\tilde{u}) - O(\varepsilon^2) + \text{Area}\ (\Sigma_2) + O(\varepsilon^3),$$

since $E(\tilde{u}|B(x_0,\varepsilon)) = O(\varepsilon^2)$, because $d\tilde{u}(x_0) \neq 0$, and the energy of $\pi^{-1} \circ w$ is the area of its image, as π and w and hence also $\pi^{-1} \circ w$ are conformal. Thus, for sufficiently small $\varepsilon > 0$, (4.5.3) is satisfied, and the proof is complete.

4.6. Nonexistence for constant boundary values

In contrast to Thm. 4.3, Lemaire [L1] showed

Proposition 4.2: There is no nonconstant harmonic map from the unit disc D onto S^2 mapping ∂D onto a single point.

Proof: Suppose $u: D \to S^2$ is harmonic with $u(\partial D) = p \in S^2$. Since the boundary values of u are constant, u is also a critical point with respect to variations $u \circ \psi$, where $\psi: D \to D$ is a diffeomorphism, mapping ∂D onto itself, but not necessarily being the identity on ∂D.

Thus, one can again use the standard argument as in 3.3 to show that u is a conformal map (cf. [L1] or [M3], pp. 369 - 372). Since u is constant on ∂D one can extend it by reflection as a conformal map on the whole of $\mathbb{R}^2$. But then this conformal map is constant on a curve interior to its domain of definition, namely ∂D, and thus has to be constant itself.

q.e.d.

The same argument was used independantly and in a different context by H. Wente [Wt].
Prop. 4.2 shows that the assumption that the boundary values are non-constant, cannot be deleted in Thm. 4.3.
By way of contrast, it is easy to see that there do exist nontrivial harmonic maps from an annulus onto a sphere with constant boundary values, namely those which map the annulus onto a geodesic loop.

<u>Remark</u>: Higher dimensional generalizations of Prop. 4.2 were obtained by Wood [W2] and Karcher - Wood [KW] (cf. also [Se2]).
Proposition 4.2 shows that the hypothesis that the boundary values are not constant is necessary for Thm. 4.3 to hold. Cor. 12.2 will give another nonexistence result, namely that there is no harmonic map at all from a torus onto a 2 - sphere with degree one. Finally, even under the assumptions of Thm. 4.3, not every homotopy class contains an energy minimizing map, as the following argument shows which is based on an idea of Lemaire [L1].
Let D be the unit disc in the complex plane, and $k: D \to S^2$ be a conformal map mapping D onto the upper hemisphere and ∂D onto the equator. Furthermore, suppose that k is equivariant with respect to the rotations of D and S^2 (the latter ones leaving the north and south pole of S^2 fixed).
We choose the orientation on S^2 in such a way that the Jacobian of k is positive.
Let $D(0,r)$ be the plane disc with center 0 and radius r (i.e. $D = D(0,1)$).
Let h_r be a map from $D(0,r)$ onto S^2 which maps $\partial D(0,r)$ onto the north pole, is injective in the interior of $D(0,r)$ and has a positive Jacobian there, and is ε - conformal. We introduce polar coordinates (ρ,φ) on D and define for $0 < r < 1$ the mapping k_r by

$$k_r(\rho,\varphi) = \begin{cases} k(\frac{1}{1-r}\rho + \frac{r}{r-1}, \varphi) & \text{if } r \leq \rho \leq 1 \\ h_r(\rho,\varphi) & \text{if } 0 \leq \rho \leq r \end{cases}$$

Using the ε-conformality theorem it is easy to see that the energy of k_r can be made arbitrarily close to 6π if we choose $r > 0$ sufficiently small.

On the other hand, 6π is just the area of the image of k_r, counted with multiplicity. Hence, if there is an energy minimizing map homotopic to k_r, its energy has to be 6π, and it therefore has to be conformal. Since the boundary values are equivariant, this conformal map itself has to be equivariant (otherwise there would exist infinitely many homotopic conformal maps with the same boundary values which is not possible). This, however, implies that it would have to collapse a circle in D to a point which is not possible for a conformal map. Hence there is no energy minimizing map homotopic to k_r.

By letting h_r cover S^2 more than once, we obtain other classes without energy minimizing maps by a similar argument. If h_r, however, has degree -1, then k_r is homotopic to a map of D onto the lower hemisphere and hence homotopic to an energy minimzing map. Hence, in this example, there are precisely two homotopy classes which contain energy minimizing maps, and they are related to each other by the operation of the proof of Thm. 4.3. A different proof of this fact was given in [BC2].

Remark: On the other hand, the proof of Thm. 4.3 shows that in many cases, we can get more than two homotopically distinct harmonic maps. Apart from the possibility that there might be several homotopy classes $\tilde{\alpha}$ with $E_{\tilde{\alpha}} = E$, our construction of the map v can yield two different new homotopy classes as soon as the functional determinant of $\tilde{u}$, a map with $E(\tilde{u}) = E$, changes sign or if there exists a point where the functional determinant of $\tilde{u}$ vanishes, while $d\tilde{u}$ is nonzero at this point.

4.7. Existence results in arbitrary dimensions

Actually, for Thm. 4.2 to hold, it is only necessary that the dimension of the domain is 2, and that the second homotopy group of the image vanishes, but not that the image is also twodimensional, cf. [L1] and [SkU]. (The proof given here also immediately generalizes to this situation).

The situation changes, however, if the dimension of the domain exceeds 2. For example, if $n \geq 3$, then for any (compact) Y and a homotopy class $\alpha \in [S^n,Y]$, the minimum of energy can be obtained in α only if α contains a constant map (which then is minimal), cf. [EL4] for more details. Therefore, in this situation, a minimizing procedure cannot yield a harmonic map in a prescribed homotopy class. The basic existence result in higher dimensions is due to Eells - Sampson [ES]. They proved the existence of a harmonic map in a prescribed homotopy class $\alpha \in [X,Y]$ (for compact X,Y) in the case where Y has nonpositive sectional curvature (Note that this is not in contrast to the difficulty noted above, since, if Y is nonpositively curved, any map $S^n \to Y$ is homotopic to a constant, since Y is a $K(\pi,1)$ manifold). They use the heat flow method, i.e. they solve the parabolic problem

$$\text{(4.7.1)} \qquad \frac{\partial u(t,x)}{\partial t} = \tau(u(t,x)) \quad \text{for } t \in (0,\infty)\ ,\ x \in X$$

$$u(0,x) = \varphi(x)$$

where φ is any smooth map in α. They show that under the curvature hypothesis on Y a solution of (4.7.1) exists for all $t \in [0,\infty)$, and that for some sequence $t_n \to \infty$, $u(t_n,\cdot)$ converges towards a harmonic map $u(\cdot) \in \alpha$.

Later on, Hartman [Ht] simplified their proof and showed that the choice of the sequence t_n is not necessary and that $u(t,\cdot)$ converges uniformly to a harmonic map as $t \to \infty$. Nevertheless, this existence result can still be regarded as fundamental.[1] It can also be proved by a minimizing procedure, using the approach of partial regularity, outlined in 4.4. Furthermore, Eells [E] observed that these partial regularity results also yield the existence of a harmonic map in a given homotopy class $\alpha \in [X,Y]$, if $\dim X = 3$, $\dim Y = 2$, $\pi_2(Y) = 0$.

The general existence problem is still to a large extent unsolved. Though one does not expect existence in all cases, so far no nonexistence results have been proved for closed solutions in higher dimensions (in contrast to the case of the Dirichlet problem, cf. the remark at the end of 4.6). In the case where the image is a $K(\pi,1)$ manifold, whence there are at least no topological obstructions, one can minimize energy in a given homotopy class. The minimum is known to be regular if there exists a strictly convex function on the universal cover of the image, using the results outlined in 4.4.

1) A different proof was given by Uhlenbeck [U].

5. Uniqueness theorems

5.1. Composition of harmonic maps with convex functions

In this section, we shall display an elementary composition property which shall be useful in the sequel.
First of all, if $u \in C^2(X,Y)$ is a map between Riemannian manifolds, and $f \in C^2(Y,\mathbb{R})$ is a function, then the following Riemannian chain rule is valid.

$$(5.1.1) \qquad \Delta(f\circ u) = D^2 f(u_{e^\alpha}, u_{e^\alpha}) + \langle (\operatorname{grad} f)\circ u\,,\ \tau(u)\rangle_Y\,,$$

where e^α is an orthonormal frame on X .
In particular, if u is harmonic, i.e. $\tau(u) = 0$, this reads as

$$(5.1.2) \qquad \Delta(f\circ u) = D^2 f(u_{e^\alpha}, u_{e^\alpha})$$

or in local coordinates

$$\Delta(f\circ u) = \gamma^{\alpha\beta} D^2 f(u_{x^\alpha}, u_{x^\beta})\,.$$

As a consequence, if f is a (strictly) convex function on Y and u is a harmonic, then $f\circ u$ is a subharmonic function on X . Actually, it is not difficult to see, that one can characterize harmonic maps by this property, as was noted by Ishihara [Ih].
We also note the following consequence (cf. Gordon [Go]).

Proposition 5.1: Suppose X is a compact manifold, possibly with boundary, and $u\colon X \to Y$ is harmonic. If there exists a strictly convex function on $u(X)$, and $u(\partial X)$ is constant in case $\partial X \neq \emptyset$, then u is a constant mapping.

Proof: From the maximum principle for subharmonic functions, it follows that $f\circ u$ is constant, and since f has definite second fundamental form, (5.1.2) implies that u itself is constant.

5.2. The uniqueness theorem of Jäger and Kaul

In this section, we want to prove the following uniqueness theorem of Jäger and Kaul for the Dirichlet problem for harmonic maps.

Theorem 5.1: Suppose that $u_i\colon \overline{\Omega} \to N$ are harmonic maps of class $C^0(\overline{\Omega},N) \cap C^2(\Omega,N)$, Ω is a bounded domain and $u_i(\overline{\Omega}) \subset B(p,M)$, where

B(p,M) is a geodesic ball in N, disjoint to the cut locus of p and with radius $M < \pi/2\kappa$ (κ^2 is an upper bound for the sectional curvature of B(p,M)).
Then the function θ,

$$\theta(x) := \frac{q_\kappa(d(u_1(x), u_2(x))}{\cos(\kappa d(p,u_1(x)))\cdot\cos(\kappa d(p,u_2(x)))}$$

$$(q_\kappa(t) := \begin{cases} \frac{1}{\kappa^2}(1-\cos\kappa t), & \text{if } \kappa > 0 \\ \frac{t^2}{2}, & \text{if } \kappa = 0 \end{cases}),$$

satisfies the maximum principle

$$\sup_\Omega \theta \leq \sup_{\partial\Omega} \theta \tag{5.2.1}$$

In particular, if $u_1|\partial\Omega = u_2|\partial\Omega$, then

$$u_1 \equiv u_2 .$$

The proof of Thm. 5.1 will actually show that we have strict inequality in (5.2.1) unless $\theta \equiv$ const. Furthermore, Thm. 5.1 also holds for weakly harmonic maps (cf. [JäK1]). The proof of Thm. 5.1 was given in [JäK2]. We shall follow the arguments of Jäger and Kaul. (The last argument will be somewhat simpliefied, however). Thus, we start with the following lemma, the proof of which is obtained by Karcher's Jacobi field technique (cf. [K2]).

Lemma 5.1: Suppose that $\gamma: [0,\rho] \to N$ is a geodesic arc parametrized by arclength, i.e. $|\gamma'| = 1$, and $0 < \rho < \pi/\kappa$. If X is a Jacobi field along γ with

$$\langle X, \gamma' \rangle = 0 , \tag{5.2.2}$$

then

$$\langle X, X' \rangle \Big|_0^\rho \geq \frac{s_\kappa'(\rho)}{s_\kappa(\rho)}(|X(0)|^2 + |X(\rho)|^2) - \frac{2}{s_\kappa(\rho)}|X(0)|\cdot|X(\rho)| \tag{5.2.3}$$

Here, we have defined

$$s_\kappa(t) := \begin{cases} \frac{1}{\kappa}\sin\kappa t & , \text{if } \kappa > 0 \\ t & , \text{if } \kappa = 0 . \end{cases}$$

Note that s_κ is the solution of

$$s_\kappa'' + \kappa^2 s_\kappa = 0 \ , \ s_\kappa(0) = 0 \ , \ s_\kappa'(0) = 1 \ .$$

Furthermore $q_\kappa(t) = \int_0^t s_\kappa$, where q_κ was defined in Thm. 5.1)

<u>Proof:</u> Let

$$s(t) := \frac{1}{s_\kappa(\rho)} \cdot (|X(0)|\, s_\kappa(\rho - t) + |X(\rho)|\, s_\kappa(t)) .$$

Then s solves

$$(5.2.4) \qquad s'' + \kappa^2 s = 0 \ , \ s(0) = |X(0)| \ , \ s(\rho) = |X(\rho)| \ ,$$

and

$$s \geqq 0 \quad \text{on } [0,\rho]$$

and

$$(5.2.5) \qquad s'(0) = \frac{1}{s_\kappa(\rho)} \, (|X(\rho)| - s_\kappa'(\rho)|X(0)|)$$

$$s'(\rho) = \frac{1}{s_\kappa(\rho)} \, (s_\kappa'(\rho)|X(\rho)| - |X(0)|) \ .$$

Then, the function

$$g := s|X|' - s'|X|$$

is differentiable, where $|X| \neq 0$. (Note that the zeros of X are isolated, since X solves the Jacobi equation

$$(5.2.6) \qquad X'' + R(X,\gamma')\gamma' = 0 \quad ,$$

which is a linear second order equation.)
Moreover,

$$g' = s|X|'' - s''|X| = s\left(\frac{\langle X,X'\rangle}{|X|}\right)' + \kappa^2 s|X|$$

$$= s\,\frac{1}{|X|^3}\,(|X|^2|X'|^2 - \langle X,X'\rangle^2) - s\cdot\frac{1}{|X|}\,\langle X,R(X,\gamma'),\gamma'\rangle + \kappa^2 s|X$$

$$\geqq 0 \ ,$$

since by assumption $\langle X, R(X,\gamma')\gamma'\rangle \leq \kappa^2 |X|^2$. Thus g is not decreasing on those intervals where it is differentiable. As was noted above, points τ where g' does not exist, i.e. $|X(\tau)| = 0$ are discrete, and moreover

$$g(\tau+0) - g(\tau-0) = 2s(\tau)|X'(\tau)| \geq 0 .$$

Thus, g is not decreasing on $[0,\rho]$, and defining

$$|X|'(\rho) = \lim_{\varepsilon \downarrow 0} |X|'(\rho-\varepsilon), \quad |X|'(0) = \lim_{\varepsilon \downarrow 0} |X|'(\varepsilon) ,$$

we conclude

$$\begin{aligned} 0 \leq g(\rho) - g(0) &= s(\rho)|X|'(\rho) - s'(\rho)|X(\rho)| - \\ &\quad - s(0)|X|'(0) + s'(0)|X(0)| = \\ &= \langle X,X'\rangle(\rho) - \langle X,X'\rangle(0) - \\ &\quad - \frac{s_\kappa'(\rho)}{s_\kappa(\rho)}\left(|X(0)|^2 + |X(\rho)|^2\right) + \\ &\quad + \frac{2}{s_\kappa(\rho)}|X(0)|\cdot|X(\rho)| , \end{aligned}$$

by (5.2.5).

q.e.d.

By assumption and 2.4, any two points $y_1, y_2 \in B(p,M)$ can be joined by a unique minimal geodesic in $B(p,M)$, and we can measure the distance between y_1 and y_2 by the length of the geodesic arc between them. We denote this (possibly modified) distance function again by $d(y_1,y_2)$. Then

$$Q_\kappa(y_1,y_2) := q_\kappa(d(y_1,y_2))$$

defines a C^2 function on $B(p,M) \times B(p,M)$, since $q_\kappa'(0) = 0$
Moreover, we note that

$$T_y(N\times N) = T_{y_1}N \oplus T_{y_2}N \quad \text{(isometrically)}$$

for $y = (y_1,y_2) \in N\times N$.
In the following lemma, we shall estimate the Hessian of Q_κ on

$B(p,M) \times B(p,M)$, using the Jacobi field estimate of Lemma 5.1

Lemma 5.2: If $y_1 \neq y_2$, then for all

$v \in T_y(N \times N)$, $y = (y_1, y_2)$, $y_1, y_2 \in B(p,M)$

$$(5.2.7) \qquad D^2 Q_\kappa(v,v) \geq \frac{\langle \operatorname{grad} Q_\kappa(y), v\rangle^2}{2Q_\kappa(y)} - \kappa^2 Q_\kappa(y)|v|^2$$

If v has the special form $0 \oplus u$ or $u \oplus 0$, then

$$(5.2.8) \qquad D^2 Q_\kappa(v,v) \geq (1 - \kappa^2 Q_\kappa(y))|u|^2 ,$$

and this also holds for $y_1 = y_2$.

Proof: First some definitions

$\rho := d(y_1, y_2)$

$v =: v_1 \oplus v_2 \in T_{y_1}N \oplus T_{y_2}N$,

$\gamma: [0,\rho] \to B(p,M)$ is the unique geodesic arc from y_1 to y_2 with $|\gamma'| = 1$,

$e_1(y) := -\gamma'(0)$

$e_2(y) := \gamma'(\rho)$

$v_i^{tan} := \langle v_i, e_i(y)\rangle\, e_i(y)$

$v_i^{nor} := v_i - v_i^{tan} \qquad (i = 1,2)$

Then, since $\rho > 0$,

$\operatorname{grad} d(y) = e_1(y) \oplus e_2(y)$,

$\operatorname{grad} Q_\kappa(y) = s_\kappa(\rho)(e_1(y) \oplus e_2(y))$, and

$D^2 Q_\kappa(y)(v,v) = \langle D_v \operatorname{grad} Q_\kappa, v\rangle$

$$(5.2.9) \qquad = s'_\kappa(\rho)\langle e_1(y) \oplus e_2(y), v_1 \oplus v_2\rangle^2 + s_\kappa(\rho) D^2 d(v,v)$$

If $\gamma_s(t)$ is the geodesic arc with

$\gamma_s(0) = \exp_{y_1}(s v_1^{nor})$, $\gamma_s(\rho) = \exp_{y_2}(s v_2^{nor})$

(note that γ_s is unique, if $s \geq 0$ is small enough), then

$$X(t) := \frac{\partial}{\partial s} \gamma_s(t)_{|s=0} \tag{5.2.10}$$

is a Jacobi field along γ with

$$X(0) = v_1^{nor} \quad , \; X(\rho) = v_2^{nor}$$

By Synge's formula (cf. [GKM], §4.1),

$$\begin{aligned} D^2 d(v,v) &= \frac{\partial^2}{\partial s^2} \text{length } (\gamma_s)_{|s=0} \\ &= \int_0^\rho (|X'|^2 - \langle X, R(X,\gamma')\gamma' \rangle)dt \end{aligned} \tag{5.2.11}$$

(note that there is no boundary term, since

$$\langle X,\gamma' \rangle = 0 \quad)$$

Since X satisfies (5.2.6), we can apply Lemma 5.1 to obtain

$$\begin{aligned} D^2 d(v,v) &= \int_0^\rho (|X'|^2 + \langle X, X'' \rangle)dt \\ &= \langle X, X' \rangle \Big|_0^\rho \geq \\ &\geq \frac{s_\kappa'(\rho)}{s_\kappa(\rho)} (|v_1^{nor}|^2 + |v_2^{nor}|^2) - \frac{2}{s_\kappa(\rho)} |v_1^{nor}| \cdot |v_2^{nor}| \quad , \end{aligned}$$

and thus with (5.29)

$$\begin{aligned} D^2 Q_\kappa(v,v) \geq\; & S_\kappa'(\rho)(\langle e_1 \oplus e_2, v_1 \oplus v_2 \rangle^2 + |v_1^{nor}|^2 + |v_2^{nor}|^2) \\ & - 2|v_1^{nor}||v_2^{nor}| \end{aligned} \tag{5.2.12}$$

If $v = 0 \oplus u$, (5.2.12) implies

$$\begin{aligned} D^2 Q_\kappa(v,v) &\geq s_\kappa'(\rho) \langle e_2(y), u \rangle^2 + s_\kappa'(\rho)|u^{nor}|^2 = \\ &= s_\kappa'(\rho)|u|^2 = \\ &= (1 - \kappa^2 Q_\kappa(y))|u|^2 \; , \end{aligned}$$

while in the general case, we only have

$$\langle e_1 \oplus e_2, v_1 \oplus v_2 \rangle^2 \leq 2(|v_1^{tan}|^2 + |v_2^{tan}|^2) ,$$

and

$$|v_i|^2 = |v_i^{tan}|^2 + |v_i^{nor}|^2 ,$$

and therefore from (5.2.12)

$$D^2 Q_\kappa(v,v) \geq s_\kappa'(\rho) \langle e_1 \oplus e_2, v_1 \oplus v_2 \rangle^2 - (1 - s_\kappa'(\rho))(|v_1^{nor}|^2 + |v_2^{nor}|^2) \geq$$

$$\geq \frac{1}{2}(1 + s_\kappa'(\rho)) \langle e_1 \oplus e_2, v_1 \oplus v_2 \rangle^2 - (1 - s_\kappa'(\rho)) (|v_1|^2 + |v_2|^2) =$$

$$= \frac{1}{2Q_\kappa(y)} \langle \operatorname{grad} Q_\kappa(y), v \rangle^2 - \kappa Q_\kappa(y)(|v_1|^2 + |v_2|^2)$$

q.e.d.

We now are in a position to prove Thm. 5.1.
We assume that θ has a positive maximum at some interior point $x_o \in \Omega$.
Then, θ is positive in a neighborhood of x_o, and $\log \theta > -\infty$ in this neighborhood.
We define

$$\psi(x) := Q_\kappa(u_1(x), u_2(x)) = \begin{cases} \frac{1}{\kappa^2}(1 - \cos \kappa d(u_1(x), u_2(x)) & \text{if } \kappa > 0 \\ \frac{1}{2} d^2(u_1(x), u_2(x)) & \text{if } \kappa = 0 \end{cases}$$

$$\varphi_i(x) := \cos(\kappa d(p, u_1(x))) , \qquad i = 1,2$$

Then $\theta = \dfrac{\psi}{\varphi_1 \cdot \varphi_2}$, and consequently

$$\text{(5.2.13)} \qquad \operatorname{grad} \log \theta = \frac{\operatorname{grad} \psi}{\psi} - \frac{\operatorname{grad} \varphi_1}{\varphi_1} - \frac{\operatorname{grad} \varphi_2}{\varphi_2} ,$$

and

$$(5.2.14)\qquad \Delta \log \theta = \frac{\Delta\psi}{\psi} - \frac{|\mathrm{grad}\ \psi|^2}{\psi^2} - \frac{\Delta\varphi_1}{\varphi_1} + \frac{|\mathrm{grad}\ \varphi_1|^2}{\varphi_1^2} - \frac{\Delta\varphi_2}{\varphi_2} + \frac{|\mathrm{grad}\ \varphi_2|^2}{\varphi_2^2}$$

Since $x \to u(x) = (u_1(x), u_2(x)) \in B(p,M) \times B(p,M)$ is also harmonic, we can make use of the chain rule (5.1.2) in order to apply Lemma 5.2. This yields

$$(5.2.15)\qquad \Delta\psi \geq \frac{|\mathrm{grad}\ \psi|^2}{2\psi} - \kappa^2\psi(|du_1|^2 + |du_2|^2),$$

since

$$|\mathrm{grad}\ \psi|^2 = \sum_\alpha \left\langle (\mathrm{grad}\ Q_\kappa)\circ u\ ,\ du(e_\alpha)\right\rangle^2 ,$$

where e_α is an orthonormal frame on Ω .
Similarly, from (5.2.8), since

$$\varphi_i(x) = 1 - \kappa^2 Q_\kappa(p, u_i(x)),$$

we obtain

$$(5.2.16)\qquad \Delta\varphi_i(x) \leq -\kappa^2\varphi_i|du_i|^2$$

Finally, by (5.2.13),

$$(5.2.17)\qquad -\frac{1}{2}\frac{|\mathrm{grad}\ \psi|^2}{\psi^2} + \frac{|\mathrm{grad}\ \varphi_1|^2}{\varphi_1^2} + \frac{|\mathrm{grad}\ \varphi_2|^2}{\varphi_2^2} \geq$$

$$\geq -\left\langle \mathrm{grad}\log\theta\ ,\ \frac{1}{2}\,\mathrm{grad}\log\theta + \frac{\mathrm{grad}\ \varphi_1}{\varphi_1} + \frac{\mathrm{grad}\ \varphi_2}{\varphi_2}\right\rangle$$

Putting

$$k(x) := \frac{1}{2}\,\mathrm{grad}\log\theta + \frac{\mathrm{grad}\ \varphi_1}{\varphi_1} + \frac{\mathrm{grad}\ \varphi_2}{\varphi_2}$$

and plugging (5.2.15), (5.2.16), and (5.2.17) into (5.2.14), we obtain

$$\Delta\log\theta + \left\langle \mathrm{grad}\log\theta\ ,\ k(x)\right\rangle \geq 0$$

Therefore $\log \theta$ is a subsolution of a linear elliptic equation, and Thm. 5.1 follows from E. Hopf's maximum principle, since the assumption that θ has a positive maximum in the interior leads to a contradiction.

5.3. Uniqueness for the Dirichlet problem, if the image has nonpositive curvature

In this section, we want to show that, if N has nonpositive sectional curvature, we can skip the assumption $u_i(\overline{\Omega}) \subset B(p,M)$ in Thm. 5.1. This result is due to Hamilton [Hm]. We shall use an observation of Schoen and Yau [SY3], which shall enable us to carry over the proof of Thm. 5.1 (those calculations in the context of Thm. 5.2 were also given by Schoen and Yau in [SY3]).

Theorem 5.2: Suppose N has nonpositive sectional curvature, and $u_i: \overline{\Omega} \to N$, $i = 1,2$, are harmonic maps, $\overline{\Omega}$ is a bounded domain, and $u_i \in C^2(\Omega,N) \cap C^0(\overline{\Omega},N)$. If $u_1|\partial\Omega = u_2|\partial\Omega$, and u_1 and u_2 are homotopic, then

$$u_1 \equiv u_2 .$$

Proof: Let $\tilde{\Omega}$ and $\tilde{N}$ be the universal covers of $\tilde{\Omega}$ and N, resp. Since N has nonpositive sectional curvature, $\tilde{d}^2$ is smooth on $\tilde{N} \times \tilde{N}$, where $\tilde{d}$ denotes the distance function of $\tilde{N} \times \tilde{N}$. $\pi(N)$ acts as a group of isometrics on $\tilde{N} \times \tilde{N}$ via

$$\alpha(x,y) = (\alpha(x),\alpha(y)) \quad \text{for } \alpha \in \pi_1(N) ,$$

and $\tilde{d}$ thus induces a function

$$\overline{d}: \tilde{N} \times \tilde{N}/\pi_1(N) \to \mathbb{R} .$$

Let $F: \overline{\Omega} \times [0,1] \to N$ be a homotopy between u_1 and u_2, with $F(x,0) = u_1(x)$ and $F(x,1) = u_2(x)$ for $x \in \overline{\Omega}$. We choose a lifting $\tilde{F}: \tilde{\Omega} \times [0,1] \to \tilde{N}$ and thus obtain liftings $\tilde{u}_1 = \tilde{F}(\cdot,0)$ and $\tilde{u}_2 = \tilde{F}(\cdot,1)$. Then for any $\gamma \in \pi_1(\Omega)$, there exists $\alpha \in \pi_1(N)$ with

$$\tilde{u}_1(\gamma(x)) = \alpha\tilde{u}_1(x) \quad \text{and} \quad \tilde{u}_2(\gamma(x)) = \alpha\tilde{u}_2(x) \tag{5.3.1}$$

for all $x \in \overline{\Omega}$.

Thus $\tilde{u}: \tilde{\Omega} \to \tilde{N} \times \tilde{N}$, defined by $\tilde{u}(x) = (\tilde{u}_1(x), \tilde{u}_2(x))$, is harmonic and by (5.3.1) induces a harmonic map

$$u : \Omega \to \tilde{N} \times \tilde{N}/\pi_1(N) .$$

Then

$$\theta := 1/2\bar{d}^2 \circ u$$

is a smooth function on Ω , and we can carry over the arguments of the proof of Thm. 5.1 to show that θ satisfies the maximum principle

$$\sup_{\Omega} \theta \leq \sup_{\partial\Omega} \theta . \tag{5.3.2}$$

q.e.d.

5.4. Uniqueness results for closed solutions, if the image has nonpositive curvature

The arguments of the preceding section also yield the following uniqueness theorem of Hartman [Ht], as was shown in [SY3].

Theorem 5.3: a) Suppose M is a compact manifold without boundary, N has negative sectional curvature, and $u: M \to N$ is harmonic. Then there is no other harmonic map homotopic to u unless $u(M)$ is contained in a geodesic of N .
b) If N only has nonpositive sectional curvature, and u_1 and u_2 are homotopic harmonic maps from M to N , then there exists a smooth family h_t , $t \in [0,1]$, of harmonic maps with $h_0 = u_1$ and $h_1 = u_2$, and the curve $h_t(x)$, $t \in [0,1]$ is for any $x \in M$ a geodesic arc, parametrized proportionally to arclength and with length independant of x .

Proof: We construct a function θ in the same way as in the proof of Thm. 5.2, assuming that u_1 and u_2 are homotopic harmonic maps. From (5.2.15) we infer again that $\theta(x) = 1/2\, d^2(u_1(x), u_2(x))$ is a subharmonic function on M and therefore constant by the maximum principle. If we put for i = 1,2 $y_i = \tilde{u}_i(x)$, $v_i = d\tilde{u}_i(e_\alpha)$, where $e_\alpha \in T_x\tilde{M}$, and $\gamma = \gamma_x$ is the geodesic between y_1 and y_2 , then we see from the proof of Thm. 5.2, that $D^2 d(v,v) = 0$, where $v = v_1 \oplus v_2 \in T_{y_1}N \oplus T_{y_2}N$. From (5.2.11) and since the sectional curvature of N is nonpositive, we see that the Jacobi field X defined in (5.2.10) is parallel (i.e. $X' = 0$) and moreover

(5.4.1) $\langle X, R(X,\gamma')\gamma' \rangle = 0$

Since θ is constant, $d(\tilde{u}_1(x), \tilde{u}_2(x)) =: \rho$ is independant of x , and we define

$$\tilde{h}_t(x) = \gamma_x(\tfrac{t}{\rho}) \qquad t \in [0,1] \text{ and } x \in \tilde{M} .$$

Since $X' = 0$ and (5.4.1) holds,

(5.4.2) $d\tilde{h}_t(x)(e_\alpha) = Y_x(t)$,

where Y_x is the Jacobi field along γ_x with $Y_x(0) = d\tilde{u}_1(e_\alpha)$ and $Y_x(\rho) = d\tilde{u}_2(e_\alpha)$. (Note that Y_x is parallel, since X is parallel).
Because $\pi_1(N)$ acts by isometries on $\tilde{N} \times \tilde{N}$, we see that for $\sigma \in \pi_1(M)$, there exists $\alpha \in \pi_1(N)$ with

$$\tilde{h}_t \circ \sigma = \alpha \circ \tilde{h}_t \qquad (t \in [0,1]).$$

Therefore, we obtain induced maps

$$h_t : M \to N \quad ,$$

and $h_0 = u_1$, $h_1 = u_2$.
From (5.4.2) and since Y_x is parallel, we infer that $dh_t(x)(e_\alpha)$ is a parallel vector field along the geodesic from $u_1(x)$ to $u_2(x)$. Thus $e(h_t)(x) = 1/2 \sum_{\alpha=1}^{n} |dh_t(x)(e_\alpha)|^2$ (where e_α is an orthonormal base of T_xM) is independant of t , and all the maps h_t have the same energy. In particular, any two homotopic harmonic maps have the same energy. Therefore, all the h_t have tc be harmonic. Namely, otherwise, there would exist a harmonic map h , homotopic to h_t and thus also to u_1 , with strictly smaller energy than h_t by the existence theorem of Eells and Sampson [ES], which is a contradiction. Thus, b) is completely proved. For a), we only have to note that if N has strictly negative sectional curvature, (5.4.1) implies that X and γ_x' are proportional. Since on the other hand $\langle X, \gamma_x' \rangle = 0$ by construction, $X = 0$, and therefore Y_x is proportional to γ_x' , for any $e_\alpha \in T_xM$. Thus, γ_x is independant of x , and if u_1 and u_2 are homotopic harmonic maps, they have to be contained in the geodesic arc $\gamma = \gamma_x$.

q.e.d.

Remark: In the twodimensional, we can replace the appeal to the theorem of Eells - Sampson which we do not prove in this book by Thm. 4.2.

5.5. Uniqueness and nonuniqueness for harmonic maps between closed surfaces

We now look more carefully at uniqueness properties of harmonic maps between closed surfaces, $u: \Sigma_1 \to \Sigma_2$. If Σ_1 is the two - sphere S^2, then u is necessarily a conformal or anticonformal branched cover (unless it is constant, which is the case if $\chi(\Sigma_2) \leq 0$, and consequently the only case of interest is $\Sigma_2 = S^2$), as we shall see in Cor. 12.1. Therefore in every homotopy class $\alpha \in [S^2,S^2]$, we obtain a family of harmonic maps, which depends on $(2 + 4 \deg \alpha)$ parameters.

If Σ_1 or Σ_2 is a flat torus T^2, then composing u with an isometry of Σ_1 or Σ_2 again gives a harmonic map.

On the other hand, a surface of genus $g > 1$ does not have any isometries homotopic to the identity, since such an isometry would be conformal with respect to the underlying conformal structure and had to coincide with the identity, since $g > 1$. Thus, if Σ_2 has genus greater than 1, nonuniqueness cannot be caused by isometries of the image. We also observe that by Kneser's Theorem (cf. Thm. 12.1), any map from S^2 or T^2 to Σ_2 is trivial in the sense that its degree is zero.

Thus, in the nontrivial case, if the genus of Σ_2 is greater than one, also isometries (or conformal automorphisms, cf. **Lemma** 6.1) of the domain cannot cause nonuniqueness of harmonic maps in a prescribed homotopy class.

Actually, in the case of maps between surfaces of genus at least two, we do not know any situation, where the harmonic map in a given homotopy class of nonzero degree is not unique.

6. A - priori $C^{1,\alpha}$- estimates

6.1. Composition of harmonic maps with conformal maps

In the case, where X is a surface, we have the following useful composition property:

Lemma 6.1: Suppose $u \in C^2(X,Y)$ is harmonic and $\dim X = 2$. If $k: M \to X$ is a conformal map between the surfaces M and X, then $u \circ k$ is again harmonic, and $E(u \circ k) = E(u)$.

Proof: We only have to note that (1.3.1), namely

$$\frac{1}{\sqrt{\gamma}} \frac{\partial}{\partial x^\alpha} (\sqrt{\gamma}\, \gamma^{\alpha\beta} \frac{\partial}{\partial x^\beta} u^i) + \gamma^{\alpha\beta} \Gamma^i_{jk} \frac{\partial}{\partial x^\alpha} u^j \frac{\partial}{\partial x^\beta} u^k = 0$$

remains valid, if we replace the metric $\gamma_{\alpha\beta}(x)$ by a conformally equivalent one, say $\lambda(x)\gamma_{\alpha\beta}(x)$, $\lambda(x) > 0$, in case dim X = 2 (cf. also (1.3.4)), and the same is true for

$$E(u) = \frac{1}{2} \int_X \gamma^{\alpha\beta} g_{ij} \frac{\partial u^i}{\partial x^\alpha} \frac{\partial u^j}{\partial x^\beta} \sqrt{\gamma}\, dx^1\, dx^2$$

q.e.d.

Cor. 6.1: On twodimensional domains, conformal maps are harmonic.

We shall later on exploit Lemma 6.1 in the following way. We shall prove a-priori estimates for harmonic maps between surfaces, first for the special case where the domain is the unit disc. Then we shall derive a-priori estimates from below for the functional determinant of univalent harmonic maps, i.e. in particular conformal maps, which will enable us to apply these special estimates to the general case, taking the existence theorem for conformal maps of chapter 3 into account.

6.2. A maximum principle

The purpose of the present chapter now is to obtain a-priori estimates for general harmonic maps with image contained in a convex ball, not only for energy minimizing ones, for which we already controlled the modulus of continuity in the proof of Thm. 4.1. Of course, we could appeal to Thm. 5.1 and 4.1 to show that under these conditions every harmonic map is energy minimizing. We have a different goal in mind, however. While in Theorem 4.1, we had to require that the boundary values possess an extension with finite energy, the $C^{1,\alpha}$-estimates obtained in this chapter for the domain D and extended to arbitrary twodimensional domains in chapter 9 will eventually enable us to prove an existence theorem without this assumption using Leray-Schauder degree theory instead of variational methods.

The following estimates all hold for image manifolds of arbitrary dimension. Some of them will be stated and proved only for twodimensional ones, for simplicity of exposition, although the proof in the general case requires only minor modifications. They also hold for domains of arbitrary dimension, but here the proofs are different, cf. e.g. [JK1].

We start with the following maximum principle which can be deduced from Jost [J1], Sperner [Sp], or Tolksdorf [Td].

Lemma 6.2: Suppose Ω is a bounded domain with Lipschitz boundary $\partial\Omega$, e.g. If $u: \Omega \to Y$ is harmonic, and $u(\Omega) \subset B(p,M)$, $M < \pi/2\kappa$, and $B(p,M)$ is a ball which is disjoint to the cut locus of p, then, if

(6.2.1) $$u(\partial\Omega) \subset B(q,M) \quad \text{for some } q \in B(p,M),$$

(6.2.2) $$\max_{x\in\Omega} d(u(x),q) < \max_{x\in\partial\Omega} d(u(x),q),$$

unless u is constant on Ω.

Proof: For any $x \in \partial\Omega$, $V: = B(u(x),M) \cap B(p,M)$ is a convex set, and by assumption, $p,q \in V$. By Lemma 2.1, we can find a geodesic arc c in V with $c(0) = p$, $c(1) = q$. Since this geodesic arc is unique in $B(p,M)$ (cf. 2.4), we infer

(6.2.3) $$d(u(x),c(t)) \leq \text{ for all } x \in \partial\Omega$$

We define

$$\omega(t): = \max_{x\in\Omega} d(u(x),c(t)) \quad \text{for } t \in [0,1].$$

$\omega(t)$ is a continuous function of t, and by assumption $\omega(0) \leq M < \pi/2\kappa$.

We now claim $\omega(t) < \pi/2\kappa$ for all $t \in [0,1]$. Indeed, if we would have $\omega(t) = \pi/2\kappa$ for some t, then, by Lemma 2.2 and (5.1.2), $d^2(u(x),c(t))$ is subharmonic on Ω, and we can apply the maximum principle for subharmonic functions, to get the contradiction

$$\frac{\pi}{2\kappa} = \max_{x\in\Omega} d(u(x),c(t)) \leq \max_{x\in\partial\Omega} d(u(x),c(t)) \leq M$$

by (6.2.3).

We conclude that in particular $\omega(1) \leq \pi/2\kappa$, and applying this time the strong maximum principle for subharmonic functions, we obtain

$$\max_{x\in\Omega} d(u(x),q) = \max_{x\in\Omega} d(u(x),c(1)) < \max_{x\in\partial\Omega} d(u(x),c(1))$$

unless u is constant on Ω.

q.e.d.

The continuity argument in the preceding proof is due to Hildebrandt - Widman [HW1]. With a similar argument and using Stampacchia's maximum principle [St], one can also prove a maximum principle for weakly harmonic maps, cf. the references mentioned above.

6.3. Interior modulus of continuity

As an application of Lemma 6.2, we obtain

Proposition 6.1: Suppose Ω is a twodimensional domain, $u: \Omega \to Y$ harmonic, $u(\Omega) \subset B(p,M)$, $M < \pi/2\kappa$, and $B(p,M)$ is a ball, which is disjoint to the cut locus of p. Then the modulus of continuity of u can be estimated in dependance only on λ (where $-\lambda^2$ is a lower bound for the curvature of Ω), $i(\Omega)$ (injectivity radius), κ, M, $E(u)$, and the modulus of continuity of the boundary values $u|\partial\Omega$.

Proof: We can use the same argument as in the proof of Thm. 4.1, using Lemma 6.2 as the appropriate maximum principle this time.

q.e.d.

6.4. Interior estimates for the energy

In the interior, we can remove the dependance on $E(u)$ in the estimate of Prop. 6.1 by virtue of the following Lemma of E. Heinz ([Hz1], Lemma 8'):

Lemma 6.3: Suppose $u: D \to Y$ is harmonic, where D is the unit disc, and $u(D) \subset B(p,M)$, where $B(p,M)$ is a ball disjoint to the cut locus of p, with $M < \pi/2\kappa$. If $B(x_0,\rho) \leq D$, then for $0 < \rho^1 < \rho$, we have the inequality

$$(6.4.1) \qquad \frac{1}{2\pi} \int_{B(x_0,\rho^1)} |du|^2 dx \leq \frac{1}{\kappa \operatorname{ctg}(\kappa M)} \cdot \frac{1}{\log \rho/\rho^1} \cdot \max_{x \in \partial B(x_0,\rho)} d(u(x),u(x_0))$$

Proof: We define

$$f(x) := 1/2\, d^2(u(x),u(x_0)) \quad .$$

Then, using Lemma 2.2b),

$$\int_{B(x_o,\rho)} \log \frac{\rho}{|x-x_o|} |du|^2 \leq \frac{1}{\kappa M \operatorname{ctg}(\kappa M)} \int_{B(x_o,\rho)} \log \frac{\rho}{|x-x_o|} \Delta f \, dx =$$

$$= \frac{1}{\kappa M \operatorname{ctg}(\kappa M)} \int_{\partial B(x_o,\rho)} (f(x_o + \rho e^{i\varphi}) - f(x_o)) d\varphi \leq$$

$$\leq \frac{2\pi}{\kappa \operatorname{ctg}(\kappa M)} \max_{x \in \partial B(x_o,\rho)} d(u(x),u(x_o)) \quad ,$$

using Lemma 2.2a).
This easily implies (6.4.1).

q.e.d.

Remarks: A corresponding statement actually holds for any domain in any dimension, cf. [JK1], Bsp. 2. Furthermore, it is even valid for weakly harmonic maps, not necessarily continuous.
At the boundary, we have to argue in a different way, since Lemma 6.3 does not pertain to the boundary.

6.5. Boundary continuity

In this section, we want to prove the boundary regularity result given in [GH] (cf. also [J1]).

Proposition 6.2: Suppose $u: D \to B(p,M)$ is harmonic, where $B(p,M)$ again is a ball with $M < \pi/2\kappa$ and disjoint to the cut locus of p. If $g = u|\partial D$ is continuous, then for every $\varepsilon > 0$ we can find some $\delta > 0$, depending on ω, κ, M, the modulus of continuity of g, and on ε, for which

(6.5.1) $\quad d(u(y),u(x_o)) \leq \varepsilon \quad$ for $y \in D \cap B(x_o,\delta)$.

If g is Hölder continuous with some exponent β, then

(6.5.2) $\quad d(u(y),u(x_o)) \leq c_\alpha |y-x_o|^\alpha \quad$ for $y \in D \cap B(x_o,\delta)$

where α and c_α depend on ω, κ, M, β, and $|g|_{C^\beta}$.

Proof: We need some definitions:

$$D(x_0,R) := D \cap B(x_0,R).$$

If $x_0 \in \partial D$, let $c: [0,1] \to B(p,M)$ be the geodesic with $c(0) = p$, $c(1) = g(x_0)$, parametrized proportionally to arclength, and

$$p_t := c(t),$$
$$v_t := d^2(u(x),p_t)$$

Furthermore, let $w_{t,R}$ be the solution of

$$\Delta w_{t,R} = 0 \quad \text{on } D(x_0,R)$$
$$w_{t,R}|\partial D(x_0,R) = v_t|\partial D(x_0,R)$$

If G_R is Green's function on $D(x_0,R)$, we derive from Green's representation formula

(6.5.3) $$\int_{D(x_0,R)} \Delta v_t(x)\, G_R(x,y)dx = w_{t,R}(y) - v_t(y)$$

From the definition of v_t and $w_{t,R}$, we have

(6.5.4) $$v_t(y) = d^2(u(y),p_t) \leq (1+t)^2M^2$$

and

(6.5.5) $$w_{t,R}(x_0) = v_t(x_0) = d^2(g(x_0),p_t) \leq (1-t)^2M^2$$

We now want to exploit that the boundary values of $w_{t,R}$ on $\partial D \cap \partial D(x_0,R)$ are given by $d^2(g(x),p_t)$, i.e. controlled by assumption. Namely, given $\varepsilon > 0$ and $R > 0, R \leq R_0$, there exists some number $r = r(\varepsilon,R)$ (depending on ε, R, M, and the modulus of continuity of $d^2(g(x),p_t)$ on $\partial D \cap \partial D(x_0,R_0)$ with the property that

(6.5.6) $$w_{t,R}(y) \leq w_{t,R}(x_0) + \varepsilon$$

for all $y \in D(x_0,r)$. This is a result from potential theory (and can be found, e.g., in [GT], Thm. 8.27).
If $d^2(g(x),p_t)$ is Hölder continuous, we even have

(6.5.7) $$w_{t,R}(y) \le w_{t,R}(x_0) + \bar{c}|y - x_0|^{2\alpha} \quad \text{for } y \in D(x_0,r),$$

where α , $\bar{c}$ depend ω , κ , M , β , and $|g|_{C^\beta}$.
We now want to apply an iteration procedure, and put

$$\bar{t}: = \frac{\pi}{2M\kappa} - 1 ,$$

$$\varepsilon: = M^2((1 - (1 - \bar{t})^2)$$

$$t_i: = i\bar{t} \quad \text{for} \quad 1 \le i \le \mu - 1$$

and $t_\mu: = 1$, where μ is the smallest integer with $\mu\bar{t} \ge 1$. Furthermore, we start with some radius $R_0 < 1$ and define

$$R_i = r(\varepsilon, R_{i-1}) \quad (i=1,\ldots\mu),$$

where r is the same r as in (6.5.6).
Then, with

$$m_i: = \max_{x\in D(x_0,R_{i-1})} (v_{t_i}(x))^{1/2} ,$$

by Lemma 2.2b, (5.1.2), (6.5.3), (6.5.6)

(6.5.8) $$2\kappa m_i \operatorname{ctg}(\kappa m_i) \cdot \int_{D(x_0,R_{i-1})} |du|^2 G_{R_{i-1}}(x,y)dx + v_{t_i}(y) \le$$

$$\le w_{t_i,R_i}(y) \le (1 - t_i)^2M^2 + \varepsilon \le M^2 \quad \text{for all } y \in D(x_0,R_i)$$

Furthermore,

$$m_1 \le \pi/2\kappa \quad \text{by (6.5.4)},$$

and if $m_i \le \pi/2\kappa$, then by (6.5.8)

$$v_{t_{i+1}} \le v_{t_i} + \bar{t}M \le \pi/2\kappa ,$$

i.e. $m_{i+1} \le \pi/2\kappa$.
Therefore, by induction,

$$m_\mu \le \pi/2\kappa ,$$

and again from (6.5.8), and (6.5.6)

$$v_1(y) \leq w_{1,R_\mu}(y) \leq w_{1,R_\mu}(x_0) + \varepsilon = \varepsilon$$

for all $y \in D(x_0, R_\mu)$.

This gives the desired estimate of the modulus of continuity at the boundary, putting $\delta = R_\mu$.

In case the boundary data are Hölder continuous, we use (6.5.7), to get

$$d(u(y),u(x_0)) \leq (v_1(y))^{1/2} \leq c|y - x_0|^\alpha .$$

q.e.d.

Remarks: The proof works in any dimension, and is even valid for weakly harmonic maps, if one approximates Green's function via mollifications.

The corresponding iteration procedure in the interior is more complicated, cf. [HJW] and [GH] .

Corollary 6.2: Under the assumptions of Prop. 6.2, for every $\varepsilon > 0$ we can calculate $\delta > 0$, depending only on ε , ω , κ , M , and the modulus of continuity of $u|\partial D$, with the property that for all x_1 , $x_2 \in D$ with $d(x_1,x_2) \leq \delta$,

$$d(u(x_1),u(x_2)) \leq \varepsilon .$$

Proof: This follows from Prop. 6.1, Prop. 6.2, and Lemma 6.3 through straightforward elementary considerations.

6.6. Interior C^1-estimates

We now prove a special case of Thm. 3.1 of [JK1] .

Theorem 6.1: Suppose $u: D \to Y$ is a harmonic map between surfaces, where D is the unit disc in the complex plane, and $u(D) \subset B(p,M)$, where $B(p,M)$ is a disc with radius $M < \pi/2\kappa$ (here $-\omega^2 \leq K \leq \kappa^2$ are curvature bounds on $B(p,M)$). Then, if $B(x_0,R) \subset D$, we have the estimate

$$(6.6.1) \qquad |du(x_0)| \leq c_0 \max_{x \in B(x_0,R)} \frac{d(u(x),u(x_0))}{R}$$

with $c_0 = c_0(\omega,\kappa,M)$.

<u>Proof</u>: The essential idea of the proof is due to E. Heinz [Hz1]. We put $R_o := R/2$ and

$$\mu := \max_{x \in B(x_o, R_o)} (R_o - d(x, x_o)) \cdot |du(x)|$$

There exists $x_1 \in B(x_o, R_o)$ with

(6.6.2) $$\mu = (R_o - d(x_1, x_o)) \cdot |du(x_1))|$$
$$d := R_o - d(x_1, x_o)$$

and

(6.6.3) $$|du(x_o)| \leq \mu / R_o$$

Furthermore,

$$\delta = \delta(\vartheta) := \max_{x \in B(x_1, d\vartheta)} d(u(x), u(x_1))$$

By Prop. 6.1 and Lemma 6.3 ($\rho = R, \rho^1 = R_o = R/2$), δ can be made as small as desired by choosing ϑ sufficiently small, in other words

$$\vartheta \leq \vartheta_o \Rightarrow \delta \leq \delta_o$$

ϑ_o depends only on δ_o, ω, κ, and M. Note that in particular ϑ_o is independant of d, since by a homothetic change of the domain which leaves $E(u)$ invariant by Lemma 6.1, we can achieve $d = 1$. We take $u(x_1)$ as the center q of the coordinates h of Lemma 2.5. For a moment, we only require $\delta_o \leq M$.
Then

(6.6.4) $$\mu / d = |du(x_1)| = |d(h \circ u)(x_1)| \leq$$

$$\leq \frac{1}{\pi d^2 \vartheta^2} \int_{\partial B(x_1, d\vartheta)} |h(u(x)) - h(u(x_1))| \, |dx| +$$

$$+ \frac{1}{2\pi} \int_{B(x_1, d\vartheta)} \frac{|\Delta h \circ u|}{d(x, x_1)} dx$$

Here, we have used an easy consequence of Green's formula, cf. e.g. "Lemma 8" in [Hz1], or Lemma 2.3 in [JK1] for a more general statement. Now

(6.6.5) $\quad |h(u(x)) - h(u(x_1))| \leq c_1 \cdot \delta$

(6.6.6) $\quad |\Delta\, h \circ u(x) \leq c_2 \cdot |du(x)|^2 \leq c_2 \dfrac{\mu^2}{d^2(1-\vartheta)^2}$,

where c_1 and c_2 can be calculated from (2.6.1) and (2.6.2).
(6.6.4) - (6.6.6) imply

$$\mu/d \leq \frac{2c_1\delta}{d\vartheta} + c_2 \frac{\mu^2\vartheta}{d(1-\vartheta)^2}$$

or, if $\vartheta \leq \vartheta_o \leq 1/2$

(6.6.7) $\quad \mu \leq a/2 \dfrac{\delta(\vartheta_o)}{\vartheta} + b/2\, \vartheta\mu^2 \quad$ for all $\vartheta \leq \vartheta_o$

As we have noted above, we can choose $\vartheta_o > 0$ so small (depending only on ω , κ , and M) that

(6.6.8) $\quad ab\delta(\vartheta) \leq ab\delta(\vartheta_o) < 1 \quad$ for all $\vartheta \leq \vartheta_o$

Then (6.6.7) implies that for $\vartheta \leq \vartheta_o$, either

(6.6.9a) $\quad \mu\vartheta \geq 1/b\ (1 + \sqrt{1-ab\delta})$

or

(6.6.9b) $\quad \mu\vartheta \leq 1/b\ (1 - \sqrt{1-ab\delta}) \quad$ for <u>all</u> $\vartheta \leq \vartheta_o$

Since μ is defined independantly of ϑ , we see, letting ϑ tend to zero that only the second possibility of (6.6.9) can hold. In particular, (6.6.9b) holds for $\vartheta = \vartheta_o$, and therefore

(6.6.10) $\quad \mu \leq \dfrac{a\delta(\vartheta_o)}{\vartheta_o}$

The result then follows from (6.6.3).

q.e.d.

<u>Remark</u>: An estimate like (6.6.1) actually holds in any dimension and for any domain, cf. [JK1], Thm. 3.1. The proof basically procceeds along the same lines, needing a much more refined geometric argument to prove a suitable version of (6.6.4), however.

)ther interior C^1 - estimates were obtained by Giaquinta - Hildebrandt [GH], Choi [Ci], and Sperner [Sp].

.7. Interior $C^{1,\alpha}$ - estimates

Corollary 6.3: Under the assumptions of Thm. 6.1,

$$(6.7.1)\qquad |u|_{C^{1,\alpha}(B(x_o,R/2))} \le c_1 \ ,$$

$$c_1 = c_1(\omega,\kappa,M,R)$$

Proof:

$$(6.7.2)\qquad \Delta u + \Gamma^i_{jk} \frac{\partial u^j}{\partial x^\alpha} \frac{\partial u^k}{\partial x^\alpha} = 0 \ ,$$

nd we take as local coordinates on a neighborhood of $u(x_1)$ (for any $x_1 \in B(x_o,R/2)$) those given from Lemma 2.5. Since the magnitude of the corresponding Christoffel symbols Γ^i_{jk} is controlled by Lemma 2.5, (6.7.1) is a direct consequence of (6.6.1) and potential theory.

q.e.d.

.8. C^1 - and $C^{1,\alpha}$ - estimates at the boundary

n this section, we shall treat the twodimensional case of Thm. 3.2 of [JK1].

heorem 6.2: Suppose $u: D \to Y$ is a harmonic map between surfaces, here D again is the unit disc, and $u(D) \subset B(p,M)$, where $B(p,M)$ is a isc with radius $M < \pi/2\kappa$. Suppose $u|\partial D = g \in C^{1,\alpha}(\partial D, B(p,M))$. hen

$$(6.8.1)\qquad |u|_{C^1(D)} \le c_2 \ ,$$

$c_2 = c_2(\omega,\kappa,M,|g|_{C^{1,\alpha}},\tau)$, where $0 < \tau < \pi/2\kappa - M$ is chosen in such a ay that $B(p,M+\tau)$ is still a disc, i.e. disjoint to the cut locus of p .

roof: By Thm. 6.1 and Prop. 6.1, it is enough to prove

$$\max_{d(x,x_o)\le R} d(u(x),u(x_o)) \le cR$$

in case $d(x_0,\partial D) = R$,

or

(6.8.2) $\quad d(u(x_2),u(x_1)) \leq cR_0$

for $d(x_0,x_2) \leq R$, $d(x_0,x_1) = R$, $x_1 \in \partial D$.
This can be achieved with the method of [HKW1] . The principal idea is to reduce the Lipschitz estimate (6.8.2) at the boundary to a Lipschitz estimate for the subharmonic function $v(x) = d^2(u(x),q)$, where q is a suitable chosen point outside $B(p,M)$. In order to guarantee, however, that v is indeed a subharmonic, we first have to reduce the domain suitably, using Prop. 6.2.
We assume w.l.o.g.

(6.8.3) $\quad \tau \leq \pi/4\kappa$,

and note that any two points in $B(p,M+\tau)$ can be joined by a unique geodesic arc inside $B(p,M+\tau)$ by Thm. 2.1.
If $u(x) \neq u(x_1)$, we continue the geodesic arc from $u(x)$ to $u(x_1)$ beyond $u(x_1)$ until coming to a point $q(x) \in B(p,M+\tau)$ with $d(q(x),u(x_1)) = \tau$. Using Prop. 6.2 and possibly shrinking R , we can find a closed subregion $D_0 \subset D$ of class C^2 with

(6.8.4) $\quad B(x_0,R) \subset D_0$

(6.8.5) $\quad B(x_1,\delta) \cap D \subset D_0 \qquad$ for some $\delta > 0$

(6.8.6) $\quad \forall x \in B(x_0,R_0)\colon\ u(D_0) \subset B(q(x),\pi/2\kappa)$

To prove (6.8.2), we take any $x_2 \in B(x_0,R)$, assume w.l.o.g. $u(x_2) \neq u(x_1)$, and put $q = q(x_2)$.
By (6.8.6), Lemma 2.2b, and (5.1.2)

$$v(x) = d^2(u(x),q)$$

is subharmonic in D_0 . As in [HKW1], proof of Lemma 3 , we compare v with the harmonic function h with the same boundary values on D_0 as v i.e.

(6.8.7) $\quad \Delta h = 0 \quad$ in D_0

$$(6.8.7) \qquad h(x) = d^2(u(x),q) \quad \text{for } x \in \partial D_o .$$

By the maximum principle

$$(6.8.8) \qquad v \leqq h \quad \text{in } D_o$$

Now

$$(6.8.9) \qquad \begin{aligned} d(u(x_1),u(x_2)) &= d(u(x_2),q) - d(u(x_1),q) \quad \text{by choice of } q \\ &\leqq 1/2\tau \ (d^2(u(x_2),q) - d^2(u(x_1),q)) \leqq \\ &\leqq 1/2\tau \ (h(x_2) - h(x_1)) \quad \text{by (6.8.7) and (6.8.8).} \end{aligned}$$

Thus, (6.8.2) is reduced to a Lipschitz bound for the harmonic function h at ∂D . This is a standard result from potential theory.

q.e.d.

<u>Corollary 6.4:</u> <u>Under the assumptions of Thm. 6.2,</u>

$$(6.8.10) \qquad |u|_{C^{1,\alpha}(D)} \leqq c_3 ,$$

<u>where</u> c_3 <u>depends on the same constants as</u> c_2 .
The proof is the same as of Cor. 6.3, taking Thm. 6.2 into account.

<u>Remark:</u> Again, corresponding statements hold in any dimension and for any regular domain, cf. [JK1]. In later sections, we shall remove the assumption that the domain is the unit disc by use of conformal maps and prove (6.8.10) for all regular twodimensional domains. Furthermore, we shall even derive $C^{2,\alpha}$ estimates (cf. Thms. 10.3, 10.4).

7. A-priori estimates from below for the functional determinant of harmonic diffeomorphisms

7.1. A Harnack inequality of E. Heinz

In this chapter, we obtain estimates from below for the functional determinant of harmonic diffeomorphisms $u: D \to \Sigma$, where D is the unit disc and Σ a surface with boundary. The assumption that the domain is D will be removed in chapter 9 .

The results from this chapter are taken from [JK1]. We start with the interior estimates which are based on the ideas of [Hz6].
First of all, we have Hilfssatz 4 of [Hz6], which we formulate as

Lemma 7.1: *Suppose* $f \in C^2(B(x_o,R),\mathbb{R})$ *is a function defined in the plane disc* $B(x_oR)$ *with* $R \leq \frac{1}{2+24a}$ *, satisfying*

$$(7.1.1) \qquad |\Delta f| \leq a(|\nabla f|^2 + |f|^2)^{1/2}$$

Furthermore, we assume $f(x_o) = 0$ *and*

$$(7.1.2) \qquad 0 < |\nabla f| \leq \gamma < \infty$$

Then

$$(7.1.3) \qquad |\nabla f(x_o)| \geq 2\gamma \exp\left(-1174\left(\log \frac{2\gamma e^8}{|\nabla f(x)|}\right)^3\right)$$

for all $x \in B(x_o,R/2)$.
Lemma 7.1 provides a Harnack type gradient estimate for solutions of the differential inequality (7.1.1) in the twodimensional case. We omit the proof since it can be found in [Hz6]. The proof is based on the representation theorem of Bers - Vekua for pseudoanalytic functions which makes it possible to reduce the assertion to a suitable Harnack inequality for holomorphic functions. We note that Lemma 7.1 is valid only in two dimensions.

7.2. Interior estimates

Lemma 7.1 together with a geometric consideration now enables us to prove a suitable analogue of Hilfssatz 5 of [Hz6] for univalent harmonic maps.

Lemma 7.2: *Suppose* $u: D \to \Sigma$ *is harmonic, and that* $J(x) \neq 0$ *throughout* $\overset{\circ}{D}$ *, where* $J(x)$ *is the functional determinant of* u *at* x *. If*

$$(7.2.1) \qquad |\nabla u(x)| \leq c$$

for $x \in B(0,1/2(1+\rho))$ *for some* $\rho < 1$ *, then*

$$(7.2.2) \qquad \log\left\{49 \log \frac{e^{10}c^2}{|J(x_1)|}\right\} \leq 3^{\frac{9+96\Lambda^2c^2}{1-\rho}} \log\left\{49 \log \frac{e^{10}c^2}{|J(x_2)|}\right\}$$

<u>for</u> $x_1, x_2 \in B(0,\rho)$, <u>if</u> $|K| \leq \Lambda^2$ (K <u>is the Gauss curvature of</u> Σ).

<u>Proof:</u> The idea consists in constructing a suitable real valued function f , with

$$|\nabla f(x)| \sim J(x) ,$$

satisfying the inequality

$$|\Delta f| \leq a|f|$$

(note that this is stronger than the assumptions of Lemma 7.1). f will be obtained by composing u with the distance from a suitable geodesic. This geometric idea is the main difference to the proof of Hilfssatz 5 in [Hz6].

We define

$$(7.2.3) \qquad \rho^* := \frac{1-\rho}{2+24\Lambda^2 c^2} .$$

(7.2.1) implies for $x \in B(x_o,\rho^*)$

$$(7.2.4) \qquad d(u(x),u(x_o)) \leq \rho^* c < \pi/4\Lambda$$

Consequently, if h measures the distance from a geodesic through $u(x_o)$, taken with a negative sign on one side, then $h \circ u$ is differentiable on $B(x_o,\rho^*)$. Using the assumptions and $|\nabla h| = 1$, we obtain for $x \in B(x_o,\rho^*)$

$$(7.2.5) \qquad 0 < \frac{|J(x)|}{c} \leq |\nabla(h \circ u)(x)| \leq c$$

We now choose the geodesic through $u(x_o)$ in such a way that also

$$(7.2.6) \qquad |\nabla(h \circ u)(x_o)| \leq |J(x_o)|^{1/2} .$$

(7.2.4), Lemma 2.4, (5.1.2), (7.2.1), and $|\nabla h| = 1$ imply

$$(7.2.7) \qquad |\Delta(h \circ u)(x)| \leq \Lambda^2 c^2 |(h \circ u)(x)| .$$

Because of (7.2.7), we can apply Lemma 7.1 to $h \circ u$, and we obtain with (7.2.5) and (7.2.6) for $x \in B(x_o,\rho^*/2)$

$$\frac{|J(x_o)|}{c^2} \geq 4 \exp\{-2348 \,(\log \frac{2c^2e^8}{|J(x)|})^3\} \quad ,$$

i.e.

$$\text{(7.2.8)} \qquad \log(49 \log \frac{e^{10}c^2}{|J(x_o)|}) \leq 3 \log(49 \log \frac{e^{10}c^2}{|J(x)|})^3 \quad .$$

Covering $B(0,\rho)$ by balls of radius $\rho^*/2$, we see from the definition (7.2.3) of ρ^* that (7.2.8) implies (7.2.2) for any $x_1, x_2 \in B(0,\rho)$.

q.e.d.

Lemma 7.3: Suppose $u: D \to \Sigma$ is a univalent harmonic map. Then $J(x) \neq 0$ throughout $\overset{o}{D}$

Proof: We can use the same argument as in the proof of Lemma 3.5.

q.e.d.

(Lemma 7.3 is due to [Hz5].)

Theorem 7.1: Suppose $u: D \to u(D)$ is a bijective harmonic map, wher

$$u(D) \subset B(p,M) \quad ,$$

$B(p,M)$ being a disc with $M < \pi/2\kappa$.
Furthermore, assume that

$$\text{(7.2.9)} \qquad \text{meas}\,(u(B(o,\sigma))) \geq \mu > 0$$

for some σ , $0 < \sigma < 1$.
Then, for $x \in B(0,r), r < 1$,

$$\text{(7.2.10)} \qquad |J(x)| \geq \delta^{-1} \, ,$$

$\delta = \delta(M,\omega,\kappa,\sigma,\mu,r)$.

Proof: By (7.2.9)

$$\int_{B(o,\sigma)} |J(x)|dx \geq \mu \quad .$$

Therefore, there exists $x^o \in B(0,\sigma)$ with

$$(7.2.11) \qquad |J(x^o)| \geq \mu/\pi$$

Putting $\rho := \max(r,\sigma)$, we have for $x \in B(0,1/2(1+\rho))$ by Thm. 6.1

$$(7.2.12) \qquad |\nabla u(x)| \leq c \quad ,$$

$c = c(M,\omega,\kappa,\rho)$.

Lemma 7.3 and (7.2.12) enable us to conclude from Lemma 7.2 for $x \in B(0,\rho)$

$$\log \{49 \log \frac{e^{10}c^2}{|J(x)|}\} \leq 3^{\frac{9+96\Lambda^2c^2}{1-\rho}} \log \{49 \log \frac{e^{10}c^2}{|J(x_o)|}\}$$

and thus

$$(7.2.13) \qquad \log \{49 \log \frac{e^{10}c^2}{|J(x)|}\} \leq 3^{\frac{9+96\Lambda^2c^2}{1-\rho}} \log \{49 \log \frac{\pi e^{10}c^2}{\mu}\}$$

by (7.2.11).

Since on the other hand by (7.2.11)

$$c^2 \geq \mu/\pi \quad ,$$

we can replace c^2 by μ/π on the left hand side of (7.2.13) to obtain the desired estimate.

q.e.d.

Corollary 7.1: *Suppose that* $u: D \to \Sigma$ *is an injective harmonic map, with* $u(D) \subset B(p,M)$, $B(p,M)$ *being a disc with* $M < \pi/2\kappa$. *Then, for any* $r \in (0,1)$ *and* $x \in B(0,r)$

$$|J(x)| \geq \delta^{-1} \quad ,$$

where

$\delta = \delta(M,\omega,\kappa,r,\text{meas } u(D),E(u))$ *and* δ *depends in addition on some kind of normalization like a 3 - point condition on the boundary or fixing the image of* 0, *or*

$$\delta = \delta(M,\omega,\kappa,r,\text{meas } u(D), |u|\partial D|_{C^\alpha}).$$

Proof: For any μ < meas $u(D)$, we can find some $\varepsilon > 0$ with the property that

(7.2.14) $$\text{meas } \{q \in u(D): d(q,\partial u(D)) \geq \varepsilon\} \geq \mu \quad .$$

Since on the other hand, we can estimate the global modulus of continuity of u on D by E(u) or $|u|\partial D|_{C^\alpha}$ (cf. sections 3.3, 6.3 - 6.5) and, by univalency of u , $\partial u(D) = u(\partial D)$, we can calculate $\sigma \in (0,1)$ with

$$d(u(x),\partial u(D)) \leq \varepsilon$$

if

$$d(x,\partial D) \leq 1 - \sigma.$$

Therefore, since u is univalent

$$u(B(0,\sigma)) \geq \{q \in u(D): d(q,\partial u(D)) \geq \varepsilon\} \quad ,$$

and by (7.2.14)

$$\text{meas } (u(B(0,\sigma)) \geq \mu \ ,$$

and Cor. 7.1 follows from Thm. 7.1.

7.3. Boundary estimates

In this section, we want to derive the boundary estimate from below for the functional determinant of [JK1].

Theorem 7.2: Suppose $u: D \to \Sigma$ is harmonic, and $u(D) \subset B(p,M)$, where $B(p,M)$ again is a disc with radius $M < \pi/2\kappa$.
Suppose that $\partial u(D) = u(\partial D)$ and that $g: = u|\partial D: \partial D \to \partial u(D)$ is a C^2 - diffeomorphism with

(7.3.1) $$0 < b \leq \left|\frac{dg(\varphi)}{d\varphi}\right| \quad \text{for all } \varphi \in \partial D \ .$$

Assume furthermore that $g(\partial D)$ is strictly convex w.r.t. $u(D)$, and that we have the following estimates for the geodesic curvature of $g(\partial D)$

(7.3.2) $\quad 0 < a_1 \leqq \kappa_g(g(\partial D)) \leqq a_2 \qquad$ for all $\varphi \in \partial D$.

Then

(7.3.3) $\quad |J(x)| \geqq \delta_1^{-1} \qquad$ for all $x \in \partial D$,

where $\delta_1 = \delta_1(\omega,\kappa,M,\tau,a_1,a_2,b,|g|_{C^{1,\alpha}})$ (τ is given in Thm. 6.2).

Proof: We define $h(q) := -d(q,\partial u(D))$ for $q \in u(D)$.
(7.3.1) and (7.3.2) imply for $q \in \partial u(D)$ (cf. (2.5.1), (2.5.2), (5.1.2))

(7.3.4) $\quad \Delta(h\circ u) \geqq a_1 b^2$

This will enable us to get a lower bound for the radial derivative of $h\circ u$ at boundary points with the argument of the boundary lemma of E. Hopf. This assertion in turn implies (7.3.3), taking (7.3.1) into account.
The constants a_2 and κ control focal points of $u(\partial D)$, and a_1 and ω then determine how long the level curves of h remain strictly convex and free of double points. Taking Cor. 6.2 into account, we can therefore find a neighborhood V_o of ∂D in D with the property that h is a C^2 function with strictly convex level curves on $u(V_o)$.
Suppose $x_o \in \partial D$.
Using now (7.3.4) and Cor. 6.4, we can choose some disc $B(x_1,r_1) \subset D$, $x_o \in \partial B(x_1,r_1)$, in such a way that

(7.3.5) $\quad \Delta(h\circ u)(x) \geqq 1/2\ a_1 b^2 \qquad$ for $x \in B(x_1,r_1)$

Defining the auxiliary function $\gamma(x)$ via

$$\gamma(x) := \frac{r_1^2}{8} a_1 b^2 \left(1 - \frac{(x-x_1)^2}{r_1^2}\right) ,$$

we have

$$\Delta\gamma(x) = -1/2\ a_1 b^2 \quad ,$$

and consequently by (7.3.5)

$$\Delta(h\circ u + \gamma)(x) \geqq 0 \quad \text{on } B(x_1,r_1).$$

Moreover,

$$(h\circ u)(x_o) + \gamma(x_o) = 0$$

and

$$(h\circ u)(x) + \gamma(x) \leq 0 \quad \text{on } \partial B(x_1,r_1),$$

since by assumption u is mapped onto the side of $\partial u(D)$, where h assumes nonpositive values, and $\gamma | \partial B(x_1,r_1) = 0$.
The maximum principle now controls the derivative of $h\circ u + \gamma$ at x_o in the direction of the outer normal, namely

$$\frac{\partial}{\partial r} (h\circ u + \gamma)(x_o) \geqq 0 ,$$

and thus

(7.3.6) $$\frac{\partial}{\partial r} (h\circ u)(x_o) \geqq \frac{r_1}{4} a_1 b^2 \quad \text{by definition of } \gamma .$$

(7.3.6) and (7.3.1) imply

(7.3.7) $$|J(x_o)| \geqq \frac{r_1}{4} a_1 b^3 =: \delta_1^{-1} .$$

q.e.d.

<u>Corollary 7.2</u>: <u>Assume</u> $u: D \to \Sigma$ <u>is an injective harmonic map,</u> <u>where</u> $u(D) \subset B(p,M)$, <u>and</u> $B(p,M)$ <u>is a disc with radius</u> $M < \pi/2\kappa$. <u>Suppose that</u> $g: = u|\partial D \in C^{1,\alpha}$, <u>and that (7.3.1) and (7.3.2) hold.</u> <u>Then for all</u> $x \in D$

(7.3.8) $$|J(x)| \geq \delta_2^{-1} ,$$

$$\delta_2 = \delta_2(\omega,\kappa,M,\tau,a_1,a_2,b,|g|_{C^{1,\alpha}}).$$

<u>Proof:</u> (7.3.8) follows from Cor. 7.1, Thm. 7.2, and Cor. 6.4.

q.e.d.

<u>7.4. Discussion of the situation in higher dimensions</u>

In chapter 9, we shall obtain estimates from below for the functional determinant of univalent harmonic maps not only on D , but on arbitrary <u>twodimensional</u> domains.

We now want to discuss the question whether one can also estimate the functional determinant of univalent harmonic maps from below in higher dimensions. First of all, the argument of the proof of Thm. 7.2 is quite general and obviously not restricted to the twodimensional case (note that we did not even have to assume that u is univalent; we needed only that u maps D onto the convex side of $u(\partial D)$).
The situation is quite different in the interior, however. The following consideration is based on an observation of Eells and Lemaire (cf. [EL1], §7, and [EL2], 7.2). For any compact M , its Albanese map

$$\alpha: M \to A(M)$$

onto its Albanese torus A(M) is harmonic. Here,

$$A(M) = H^*/\Gamma \ ,$$

where H^* is the conjugate space of the real vector space of harmonic 1 - forms, and Γ is a suitable homomorphic image of $\pi_1(M)$. Thus, A(M) is a flat torus. Furthermore

$$J(\alpha) = \omega_1 \wedge \ldots \wedge \omega_n \quad ,$$

where $(\omega_i)_{i=1,..n}$ is a basis for the harmonic one - forms. By an investigation of Calabi [C], for $n \geq 3$, there are metrics g on the n - torus T^n , for which some ω_i have zeroes. Thus, also $J(\alpha)$ has zeroes for

$$\alpha: (T^n, g) \to A(T^n, g)$$

We denote the metric of $A(T^n,g)$ by g_o , and construct a smooth family g(t) of metrics with $g(0) = g_o$ and $g(1) = g$. By the existence theorem of Eells - Sampson we can find harmonic maps

$$u_t: (T^n, g(t)) \to (T^n, g_o) \quad .$$

Furthermore, any two homotopic harmonic maps u_t and v_t from $(T^n,g(t))$ onto $(T^n,g\ )$ differ only by an isometry of (T^n,g_o) by Thm. 5.3. Since the image is flat, we can derive uniform a - priori $C^{1,\alpha}$ estimates for the maps u_t , independant of t , just using the estimates for solutions of the Laplace - Beltrami equation on $(T^n,g(t))$.
Now the set of $t \in [0,1]$ for which $J(u_t) > 0$ on T^n is open and not empty, since u_o is an isometry. If we would have a lower bound for the

functional determinant of univalent harmonic maps with small image, defined on some subset of $(T^n, g(t))$, then the uniform C^α-estimate for u_t would enable us to obtain lower bounds for the functional determinant of those u_t with $J(u_t) > 0$[1]. Consequently the set of t with $J(u_t) > 0$ on T^n would also be closed, and therefore coincide with [0,1], and in particular

$$J(u_1) > 0 .$$

On the other hand, again by uniqueness, u_1 differs from α only by an isometry of (T^n, g_0), which is a contradiction, since $J(\alpha)$ has zeroes.

8. The existence of harmonic diffeomorphisms which solve a Dirichlet problem

8.1. Proof of the existence theorem in case the image is contained in a convex ball and bounded by a convex curve

We can now easily prove the main result of [J3].

Theorem 8.1: Suppose Ω is a compact domain with Lipschitz boundary $\partial\Omega$ on some surface, and that Σ is another surface. We assume that $\psi: \overline{\Omega} \to \Sigma$ maps $\overline{\Omega}$ homeomorphically onto its image, that $\psi(\partial\Omega)$ is contained in some disc $B(p,M)$ with radius $M < \pi/2\kappa$ (where $\kappa^2 \geq 0$ is an upper curvature bound on $B(p,M)$) and that the curves $\psi(\partial\Omega)$ are of Lipschitz class and convex w.r.t. $\psi(\Omega)$.
Then there exists a harmonic mapping $u: \Omega \to B(p,M)$ with the boundary values prescribed by ψ, which is a homeomorphism between $\overline{\Omega}$ and its image, and a diffeomorphism in the interior.
Moreover, if $\psi|\partial\Omega$ is even a C^2-diffeomorphism between C^2-curves, then u is a diffeomorphism up to the boundary.

Theorem 8.1 and the uniqueness theorem of Jäger and Kaul (cf. Thm. 5.1 imply

Corollary 8.1: Under the assumptions of Thm. 8.1, each harmonic map which solves the Dirichlet problem defined by ψ and which maps Ω into a geodesic disc $B(p,M)$ with radius $M < \pi/2\kappa$, is a diffeomorphism

---

1) This patching together of local estimates works as follows. By uniform continuity, for any $\varepsilon > 0$ there is some $\delta > 0$ with $u_t(B(x,\delta)) \subset B(u(x),\varepsilon)$. Thus, $u_t|B(x,\delta)$ has arbitrarily small image, and if we have estimates for maps with small image, we have corresponding ones for u_t .

in Ω .

Proof of Thm. 8.1: First of all, $\partial\Omega$ is connected. Otherwise, $\psi(\partial\Omega)$ would consist of at least two curves, both of them convex w.r.t. $\psi(\Omega)$. Therefore, we could find a nontrivial closed geodesic γ in $\psi(\Omega) \subset B(p,M)$ with an easy Arzela - Ascoli argument (cf. Lemma 2.1 for a similar statement). Since a geodesic can be considered as a special case of a harmonic map and $M < \pi/2\kappa$, Lemma 2.2.b and Prop. 5.1 imply that γ has to be a point, which is a contradiction. Therefore, $\partial\Omega$ is connected, and since Ω is homeomorphic to $\psi(\Omega)$, we conclude that Ω is a disc, topologically.

Therefore, we have to prove the theorem only for the case where Ω is the plane unit disc D , taking the existence theorem 3.1 for a conformal map $k: D \to \Omega$ and the composition property Lemma 6.1 into account.

For the rest of this section, we assume that $\psi: \partial D \to \psi(\Omega D)$ is a C^2 - diffeomorphism between curves of class $C^{2,\alpha}$, that $\psi(\partial\Omega)$ is not only convex, but strictly convex, and that we have the following quantitative bounds

(8.1.1) $$\left|\frac{d^2}{d\varphi^2}\,\psi(\varphi)\right| \leq b_1$$

and for $\varphi \in \partial D$

(8.1.2) $$\left|\frac{d}{d\varphi}\,\psi(\varphi)\right| \geq b_2^{-1}$$

and

(8.1.3) $$0 < a_1 \leq \kappa_g(\psi(\partial D)) \leq a_2 \quad .$$

These assumptions can be removed later on by approximation arguments which we shall indicate in section 8.2.

By virtue of Thm. 3.1, there is a conformal map $k: D \to \psi(D)$. By a variation of boundary values, we now want to deform this conformal map into a diffeomorphic harmonic map u .

Without loss of generality, we may assume that the boundary value map ψ preserves the orientation of ∂D . Now let γ be the parametrization of the boundary curve of $\psi(D)$ by arclength. We set

(8.1.4) $$\omega(\phi,\lambda) := \gamma(\lambda\gamma^{-1}(k(\phi)) + (1-\lambda)\gamma^{-1}(\psi(\phi))),$$

$$\phi \in \partial D \; , \; \lambda \in [0,1] \; .$$

ω deformes the boundary values of k into the boundary values prescribed by ψ .
Since we assumed that (8.1.1) and (8.1.2) hold and that $\psi(\partial D) \in C^{2,\alpha}$, application of Thm. 3.1 yields that

$$\text{(8.1.5)} \qquad \omega(\phi,\lambda) \ , \ \frac{\partial}{\partial\phi}\omega(\phi,\lambda) \quad \text{and} \quad \frac{\partial^2}{\partial\phi^2}\omega(\phi,\lambda)$$

are continuous functions of λ ,

$$\text{(8.1.6)} \qquad \frac{\partial}{\partial\phi}\omega(\phi,\lambda) \qquad \text{does not vanish for any } \phi \in \partial D \text{ and } \lambda \in [0,1].$$

Let now u_λ denote the harmonic map from D to B(p,M) with boundary values $\omega(\cdot,\lambda)$, (the existence of u_λ follows from Thm. 4.1) and let $\lambda_n \in [0,1]$ be a sequence converging to some $\lambda \in [0,1]$.
By Cor. 6.4, the Arzela-Ascoli Theorem and the uniqueness theorem 5.1, u_{λ_n} converges to the harmonic map u_λ in the $C^{1,\beta}$'- topology, $0 < \beta < \alpha$
In particular,

$$p(\lambda) := \inf_{x\in D} |J(u_\lambda)(x)|$$

depends continuously on λ ($J(u_\lambda)$ denotes the Jacobian of u_λ).
We define $L := \{\lambda \in [0,1]: p(\lambda) > 0\}$. By Thm. 3.1, $0 \in L$ (u_0 is the conformal map k), and therefore L is not empty. Since we assumed (8.1.2) and (8.1.3), which implied (8.1.5) and (8.1.6) we can apply Cor. 7.2 to the extent that

$$\text{(8.1.7)} \qquad p(\lambda) \geqq p_0 > 0 \qquad \text{for } \lambda \in L$$

Since $p(\lambda)$ depends continuously on λ , (8.1.7) implies $L = [0,1]$.
Thus, u_1 is a local diffeomorphism and a diffeomorphism between the boundaries of D and $u_1(D)$, and consequently a global diffeomorphism by the homotopy lifting theorem.
Thus, the proof of Thm. 8.1 is complete, except for the approximation arguments described in the next section.

8.2. Approximation arguments

In section 8.1, we have assumed that the boundary of the image is strictly convex, and, in addition, that the boundary values are a diffeomorphism of class C^2 . In this section, we shall prove the theorem also for the case that the boundary is only supposed to be convex and that the boundary values are only supposed to induce a homeomorphism of the boundaries.

We shall present only the first approximation argument. It is a modification of the corresponding one given by E. Heinz in [Hz4], pp. 178 - 183. The reasoning for the second case can be taken over from [Hz3], pp. 351 - 352, in case $\partial\psi(D) \in C^{2,\alpha}$. The case of a general boundary is handled by an approximation by smooth curves.

Therefore, let us suppose that the boundary of the image $\psi(D)$ is only convex, while the boundary values ψ are still assumed to be a diffeomorphism of class C^2. Then we argue in the following way:

Given a metric g_{ij} on the image with respect to which the boundary of $A := \psi(D)$ is convex, there is a sequence $\{g^n_{ij}\}$ of metrics on A such that ∂A is even strictly convex with respect to g^n_{ij}, according to [Hz4], §4. Moreover, $\{g^n_{ij}\}$ can be chosen to converge uniformly to g_{ij} on A together with their first and second derivatives, as $n \to \infty$.

Keeping the boundary values ψ fixed, we consider the map $u_n(x)$ which is harmonic in the metric g^n_{ij} and which solves the Dirichlet problem with boundary values ψ. The existence of u_n is guaranteed by the arguments given above - at least for large values of n, when g^n_{ij} is so close to g_{ij} that the geometric conditions are satisfied.

By virtue of Cor. 7.1, on each disc $B(O,r)$, $r < 1$, there is an a-priori bound of the functional determinant of $u_n(x)$ from below. Moreover, by virtue of Cor. 6.4, we can choose a subsequence of the functions $u_n(x)$ which converges uniformly on D together with the first derivatives to a map $u(x)$. In particular, the u_n converge to u strongly in H^1_2. Therefore, u is a weakly harmonic map w.r.t. the metric g_{ij}, i.e. a weak solution of the corresponding Euler equations. Since u is also of class C^1, linear elliptic regularity theory implies that u is a classical solution, i.e. harmonic. Moreover, u is a local diffeomorphism in the interior, and since it is the uniform limit of diffeomorphisms, it is a diffeomorphism in the interior.

q.e.d.

Remarks: 1) Actually, using a further approximation argument, we don't even have to assume that the boundary values are homeomorphic. We need only that they are continuous and monotonic, i.e. a uniform limit of homeomorphisms. The corresponding harmonic solution of the

Dirichlet problem still remains a diffeomorphism in the interior.

2) In the next chapter, we shall use a degree argument to obtain a non-variational proof of Thm. 8.1, i.e. one not involving Thm. 4.1.

8.3. Remarks: Plane domains, necessity of the hypotheses of Theorem 8.1

1) In the case where both Ω and $\psi(\Omega)$ are bounded simply connected domains in the plane, the assertion of Thm. 8.1 was already obtained by Radó and Kneser [Rd], [Kn1] and Choquet [Cq]. Choquet also showed that the convexity of the boundary of the image is necessary for Thm. 8.1 to hold. The reason is the following. Suppose the image has the following shape.

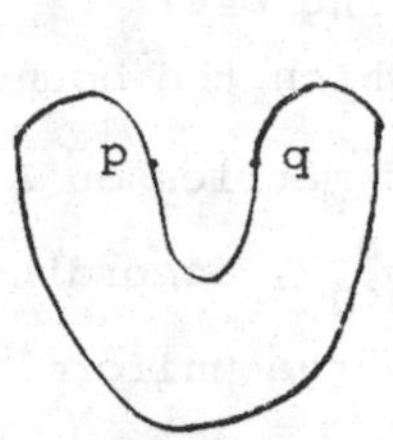

If the boundary values $\psi(\partial\Omega)$ are concentrated near p and q, then by the mean value property of harmonic functions, some points of Ω will be mapped onto points between p and q not belonging to $\psi(\Omega)$.

This is in essential contrast to the case of conformal maps where convexity of the image is not necessary to guarantee that the solution is a diffeomorphism (cf. Thm. 3.1). Note that a conformal map is a solution of a free boundary value problem instead of a Dirichlet problem.

2) The size restriction on the image in Thm. 8.1 will be removed in chapter 11 (cf. Thm. 11.2).

9. $C^{1,\alpha}$-a-priori estimates for arbitrary domains. Non-variational existence proofs

9.1. $C^{1,\alpha}$-estimates on arbitrary surfaces

In this section, we shall prove Corollaries 6.3 and 6.4 for arbitrary domains (cf. [JK1].)

Theorem 9.1: Suppose $u: \Omega \to \Sigma$ is a harmonic map between surfaces. Suppose again $u(\Omega) \subset B(p,M)$, where $B(p,M)$ is a disc with radius $M < \pi/2\kappa$ and $-\omega^2 \leq K \leq \kappa^2$ are curvature bounds on $B(p,M)$, while $-\lambda^2 \leq K_\Omega \leq \Lambda^2$ are curvature bounds on Ω.
Then for every $\Omega_0 \subset\subset \Omega$

$$(9.1.1) \qquad |u|_{C^{1,\alpha}(\Omega_0)} \leq c_{10} \quad ,$$

$$c_{10} = c_{10}(\omega,\kappa,M,\lambda,\Lambda,i(\Omega),\mathrm{diam}\,\Omega,d(\Omega_0,\partial\Omega)).$$

Proof: First of all, there is a finite number of points x_i on Ω, with the property that all $B(x_i, 2R)$ are discs and that the smaller discs $B(x_i,R)$ cover Ω_o, where $R = \min\ (1/2\ i(\Omega); 1/2\ d(\Omega_o, \partial\Omega))$. For each i, there is a conformal representation k_i of $B(x_i, 2R)$,

$$k_i\colon D \to B(x_i, 2R) \quad \text{(normalized e.g. by } k_i(0) = x_i)$$

by Thm. 3.1. Thm. 6.1 then implies (using Lemma 6.1),

$$(9.1.2) \qquad |d(u \circ k_i)|_{B(0,\rho)} \leq c_{11}(\omega, \kappa, M, \rho)$$

for every $\rho < 1$.
Moreover, since k_i is a diffeomorphism, the Courant - Lebesgue Lemma 3.1 directly bounds the modulus of continuity of k_i by $E(k_i)$ (cf. 3.3). On the other hand, $E(k_i)$ is bounded by the geometry of Ω (actually in terms of λ, Λ, R, since k_i is minimizing among diffeomorphisms (cf. proof of Thm. 3.1)). Thus, we can calculate some $\rho < 1$, for which

$$(9.1.3) \qquad k_i(B(0,\rho)) \supset B(x_i, R)$$

as in the proof of Cor. 7.1.
Furthermore

$$(9.1.4) \qquad J(k_i) \geq \delta^{-1} \qquad \text{on } B(0,\rho),$$

$\delta = \delta(\lambda, \Lambda, R, \rho)$ (cf. Thm. 7.1).
(9.1.2) - (9.1.4) then imply

$$(9.1.5) \qquad |du|_{B(x_i,R)} \leq c_{12}(\omega, \kappa, M, \lambda, \Lambda, R).$$

If we now introduce the coordinates given from Lemma 2.5 on small balls of domain and image, (9.1.1) follows from linear elliptic theory, since the equations for harmonic maps in local coordinates take the form

$$\frac{1}{\sqrt{\gamma}} \frac{\partial}{\partial x^\alpha} \left(\sqrt{\gamma}\, \gamma^{\alpha\beta} \frac{\partial}{\partial x^\beta} u^i\right) + \gamma^{\alpha\beta}\, \Gamma^i_{jk} \frac{\partial u^j}{\partial x^\alpha} \frac{\partial u^k}{\partial x^\beta} = 0 .$$

Theorem 9.2: Suppose $u\colon \Omega \to \Sigma$ is a harmonic map between surfaces, $u(\Omega) \subset B(p,M)$. Suppose $B(p, M+\tau)$, $\tau > 0$, is a disc with $M+\tau < \pi/2\kappa$, where we assume curvature bounds $-\omega^2 \leq K \leq \kappa^2$ on $B(p,M+\tau)$.

Furthermore, let $-\lambda^2 \leq K_\Omega \leq \Lambda^2$ be curvature bounds on a neighborhood of Ω, and suppose $\partial\Omega \in C^{1,\alpha}$, $g := u|\partial\Omega \in C^{1,\alpha}$.
Then

$$|u|_{C^{1,\alpha}(\Omega)} \leq c_{14} ,$$

$$c_{14} = c_{14}(\omega,\kappa,M,\tau,\lambda,\Lambda,\text{diam }\Omega,i(\Omega),|g|_{C^{1,\alpha}},|\partial\Omega|_{C^{1,\alpha}}).$$

The proof proceeds along the same lines as the one of Thm. 9.1 and is therefore omitted, in order to avoid repetition.

Remark: Thms. 9.1 and 9.2 will be considerably improved in 10.5, where we shall derive even $C^{2,\alpha}$ a-priori bounds which depend on the same geometric quantities as the $C^{1,\alpha}$ bounds!

9.2. Estimates for the functional determinant on arbitrary surfaces

Cor. 7.1 and 7.2 can be proved for arbitrary domains as well.

Theorem 9.3: Suppose $u: \Omega \to \Sigma$ is an injective harmonic map between surfaces, $u(\Omega) \subset B(p,M)$, where $B(p,M)$ is a disc with radius $M < \pi/2\kappa$. Then for any $\Omega_o \subset\subset \Omega$ and $x \in \Omega_o$,

$$(9.2.1) \qquad |J(u)| \geq \delta_3^{-1} > 0 \quad \text{in } \Omega_o ,$$

$\delta_3 = \delta_3(M,\omega,\kappa,\text{meas } u(\Omega), i(\Omega),\lambda,\Lambda,\text{diam }\Omega, d(\Omega_o,\partial\Omega), E(u))$, ($-\lambda^2 \leq K_\Omega \leq \Lambda^2$ for the curvature of Ω), and the dependance on $E(u)$ can be replaced by $|u|\partial\Omega|_{C^\alpha}$.
If moreover $g := u|\partial\Omega \in C^{1,\alpha}$, and

$$(9.2.2) \qquad 0 < b \leq \left|\frac{dg(\varphi)}{d\varphi}\right| \qquad \text{for all } \varphi \in \partial\Omega$$

and

$$(9.2.3) \qquad 0 < a_1 \leq \kappa_g(g(\partial\Omega)) \leq a_2 ,$$

then

$$|J(u)| \geq \delta_4^{-1} > 0 \qquad \text{in } \Omega$$

where δ_4 depends on the same quantities as δ_3 and on $|g|_{C^{1,\alpha}}$, a_1, a_2

b, and τ , where τ is given in Thm. 6.2 (also $-\lambda^2 \leq K_\Omega \leq \Lambda^2$ are curvature bounds in a neighborhood of Ω).

Proof: We again compose u with suitable conformal maps k_i as in the proof of Thm. 9.1. Cor. 7.1 and 7.2 bound $|J(u \circ k_i)|$ from below. Since $J(u \circ k_i) = J(u) \cdot J(k_i)$, the assertions follow, since Thm. 6.1 provides interior gradient bounds for k_i .

q.e.d.

9.3. A non - variational proof of Theorem 4.1

We are now ready to give a non - variational proof of Thm. 4.1, using Thms. 9.1 and 9.2. (Actually, there is still some variational procedure involved through the existence theorem 3.1 for a conformal map which was used in the proofs of Thm. 9.1 and 9.2. One can dispense of the use of conformal maps, in those proofs, however, by use of a more refined geometric argument as in [JK1]).

The main advantage of the present proof, compared to the one given in section 4, however, is that it is selfcontained. While in chapter 4, we had to appeal to regularity results given e.g. in [LU], in the present situation, we can use the a - priori estimates of Thms. 9.1 and 9.2 together with Leray - Schauder degree theory.

Actually, we shall prove a slightly stronger statement than Thm. 4.1, because we don't have to require that the boundary values admit an extension with finite energy.

Theorem 9.4: Suppose Ω is a bounded domain on a surface, $g: \partial\Omega \to \Sigma$ is continuous, $g(\partial\Omega) \subset B(p,M)$, where $B(p,M)$ is a disc on the surface Σ with radius $M < \pi/2\kappa$. Then there exists a harmonic map $h: \Omega \to B(p,M)$ with $h|\partial\Omega = g$.

Proof: We can assume w.l.o.g. $g \in C^2$, since the general case follows by an approximation argument as in 8.2, involving Cor. 6.2.

We want to use the Leray - Schauder degree theory as in [HKW1] and [HKW2] .

Therefore, we define

$$(9.3.1) \qquad g(\varphi,t) = \begin{cases} 0 & \text{for } 0 \leq t \leq 1/2 \\ (2t-1)g(\varphi) & \text{for } 1/2 \leq t \leq 1 \end{cases}$$

($\varphi \in \partial\Omega$),

We cover $B(p,M_o)$ by normal coordinates G centered at p, where $M < M_o < \pi/2\kappa$, and M_o is chosen in such a way that $B(p,M)$ is still a disc. We denote the corresponding Christoffel symbols by Γ^i_{jk}, and define

$$\Gamma^i_{jk}(q,t) := \begin{cases} 2t\ \Gamma^i_{jk}(q) & \text{for } 0 \leq t \leq 1/2 \\ \Gamma^i_{jk}(q) & \text{for } 1/2 \leq t \leq 1 \end{cases} \tag{9.3.2}$$

$(q \in G(B(p,M)),\ i\ ,\ j\ ,\ k = 1,2)$

If we extend Γ^i_{jk} as continuous and bounded functions from $G(B(p,M_o))$ to the whole plane $\mathbb{R}^2$, we can define the following map of $C^1(\Omega,\mathbb{R}^2)$ into itself:

$\overline{u} \to F(\overline{u},t)$,

where $v = F(\overline{u},t)$ is the solution of the Dirichlet problem

$$\begin{aligned} &\Delta v^i(x) + \gamma^{\alpha\beta}\ \Gamma^i_{jk}(\overline{u}(x))\ \frac{\partial \overline{u}^j(x)}{\partial x^\alpha}\ \frac{\partial \overline{u}^k(x)}{\partial x^\beta} = 0 \quad \text{on} \\ &v|\partial\Omega = g(\cdot,t), \end{aligned} \tag{9.3.3}$$

where $(\gamma^{\alpha\beta})$ is the metric of Ω in local coordinates and Δ is the corresponding Laplace - Beltrami operator.

<u>Lemma 9.1:</u> a) <u>For each</u> $t \in [0,1]$, $F(\cdot,t): \overline{u} \to F(\overline{u},t)$ <u>is a completely continuous transformation of</u> $C^1(\Omega,\mathbb{R}^2)$ <u>into itself.</u>
b) <u>On every ball</u> $\{|\overline{u}|_{C^1} \leq R\}$, F <u>is uniformly continuous in</u> t.

<u>Proof:</u> A consequence of linear elliptic theory.

q.e.d.

Lemma 9.1 implies that $\overline{u} \to \Phi(\overline{u},t) := \overline{u} - F(\overline{u},t)$ is of Leray - Schauder type on bounded open subsets Y of $C^1(\Omega,\mathbb{R}^2)$. Thus, we can define a degree $\deg(\Phi(\overline{u},t),Y,0)$ with respect to the image value 0 (cf. [LS]). We now define a suitable set Y.

$Y := \{\overline{u} \in C^1(\Omega,\mathbb{R}^2)$, $\overline{u}$ is the coordinate representation of some map $u: \Omega \to B(p,M)$ with

$$|\overline{u}(x)| = d(u(x),p) < M_o \quad \text{in } \Omega \text{ and} \tag{9.3.4}$$

$$|du| < 2c_{14}\} \tag{9.3.5}$$

c_{14} is the constant in Thm. 9.2, for the boundary values (9.3.1) and the metrics with Christoffel symbols (9.3.2) (note that c_{14} can be chosen independantly of t)

Lemma 9.2:

$$\deg(\Phi(\overline{u},t),Y,0) = 1 \qquad \text{for all } t \in [0,1] .$$

In particular, there is a solution of

(9.3.6) $\qquad \Phi(\overline{u},1) = 0 \quad \text{in } Y .$

Lemma 9.2 implies Thm. 9.4, since the solution of (9.3.6) is the coordinate representation of a harmonic map $h: \Omega \to B(p,M_o)$, and Lemma 6.2 then implies $h(\Omega) \subset B(p,M)$.

Proof of Lemma 9.2: Since $\Phi(\overline{u},0) = \overline{u}$ and $0 \in Y$, we see that $\deg(\Phi(\overline{u},0),Y,0) = 1$.
Thus, we only have to show that $\deg(\Phi(\overline{u},t),Y,0)$ is independant of t.
The fundamental theorem of Leray - Schauder ([LS], p. 63) then implies that $\Phi(\overline{u},1) = 0$ has a solution in Y.
In order to show that $\deg(\Phi(\overline{u},t),Y,0)$ is independant of t, we only have to exclude that

(9.3.7) $\qquad \Phi(\overline{u},t) = 0$

has a solution $\overline{u} \in \partial Y$.
If $\overline{u} \in \overline{Y}$ is a solution of (9.3.7) for some $t \leq 1/2$, then it is easily seen from Lemma 2.2 and (5.1.2) that $|\overline{u}|^2$ is subharmonic on Ω with vanishing boundary values, and consequently $\overline{u} \equiv 0$.
If $\overline{u} \in \overline{Y}$ is a solution of (9.3.7) for some $t \geq 1/2$, then $\overline{u}$ represents a harmonic map with boundary values $g(\cdot,t)$. Lemma 6.2 and Thm. 9.2 then imply $\overline{u} \in Y$.
Thus, in any case, (9.3.7) cannot have a solution $\overline{u} \in \partial Y$.

q.e.d.

9.4. A non - variational proof of Theorem 8.1

In this section, we want to provide a degree theoretic proof of Thm. 8.1. It proceeds as in [J3] and is based on arguments of E. Heinz (cf. [Hz 3]). Let us first formulate the assertion again

Theorem 9.5: Suppose under the assumptions of Thm. 9.4, that the boundary values extend to a homeomorphism ψ of $\bar{\Omega}$ onto its image, and that $\psi(\partial\Omega)$ is convex w.r.t. $\psi(\Omega)$.
Then we can find a harmonic solution of the corresponding Dirichlet problem which is a homeomorphism on $\bar{\Omega}$ and a diffeomorphism in the interior.

Proof: As in the proof of Thm. 8.1, we see that Ω is topologically a disc. We can therefore again assume that (8.1.1), (8.1.2) and (8.1.3) hold for $\varphi \in \partial\Omega$, since these assumptions can be removed by the approximation arguments of 8.2. Finally, by Thm. 3.1, there exist conformal maps $k_1 : D \to \Omega$ and $k_2 : D \to \psi(\Omega)$, and therefore also a conformal map $k := k_2 \circ k_1^{-1} : \Omega \to \psi(\Omega)$. We can again deform the boundary values of k into $\psi|\partial\Omega$ as in 8.1, and (8.1.5) and (8.1.6) pertain.
As in the proof of Thm. 9.3, we represent B(p,M) by normal coordinates We put $B := G(\psi(\Omega))$, and let again $\bar{u}$ denote the coordinate representation of u.

Definition 9.1: Let λ be an element of [0,1]. Then $K(\lambda)$ is the set of all functions $\bar{u} = \bar{u}(x)$, $\bar{u} : \Omega \to B$, with the following properties:

a) $\bar{u}(x) \in C^2(\Omega, B) \cap C^0(\bar{\Omega}, \bar{B})$, and $\bar{u}(x)$ maps Ω homeomorphically onto B with nonvanishing functional determinant,

b) u(x) is harmonic in Ω, i.e. $\bar{u} = \bar{u}(x)$ is a solution of the system

$$(9.4.1) \qquad \Delta\bar{u}^i + \gamma^{\alpha\beta}\Gamma^i_{jk}(\bar{u}) D_\alpha \bar{u}^j D_\beta \bar{u}^k = 0 \qquad (i = 1,2)$$

c) $u(\phi) = \omega(\phi, \lambda)$ $\qquad (\phi \in \partial\Omega)$.

Let $K := \bigcup_{\lambda \in [0,1]} K(\lambda)$

We now define the transformation $\bar{u} \to H(\bar{u}, \lambda)$ by the requirement that $H(\bar{u}, \lambda)$ is the solution v of

$$(9.4.2) \qquad \Delta v^i(x) + \gamma^{\alpha\beta}\,\Gamma^i_{jk}(\bar{u}(x)) \frac{\partial \bar{u}^j(x)}{\partial x^\alpha} \frac{\partial \bar{u}^k(x)}{\partial x^\beta} = 0 \qquad \text{on } \Omega$$

$$v|\partial\Omega = G(\omega(\cdot, \lambda))$$

By elliptic regularity theory

<u>Lemma 9.3:</u> <u>A function</u> $\bar{u} \in C^1(\Omega,\mathbb{R}^2)$ <u>is contained in</u> $K(\lambda)$, <u>if and only if</u> $\bar{u} = H(\bar{u},\lambda)$ <u>and the functional determinant of</u> $\bar{u}$ <u>vanishes nowhere in</u> Ω .

We now define

$$Y_o : = \{\bar{u} \in C^1(\bar{\Omega},\mathbb{R}^2),\ \bar{u} \text{ represents } u \text{ with}$$

(9.4.3) $\quad |\bar{u}(x)| = d(u(x),p) < M_o$

(9.4.4) $\quad |du| < 2c_{14} \quad$ in $\bar{\Omega}$

(9.4.5) $\quad |J(u)| > 1/2\delta_3^{-1}\}$

Here, c_{14} is defined in Thm. 9.2, and δ_3 in Thm. 9.3 (note that c_{14} and δ_3 can be chosen uniformly for the family of boundary values $g(\cdot,\lambda)$).

<u>Lemma 9.4:</u> <u>The transformation</u> $\bar{u} \to \Psi(\bar{u},\lambda) = \bar{u} - H(\bar{u},\lambda)$ of $C^1(\Omega,\mathbb{R}^2)$ <u>into itself is of Leray-Schauder type.</u>

<u>Moreover,</u>

(9.4.6) $\quad \deg(\Psi(\bar{u},\lambda),Y_o,0) = 1 \qquad$ <u>for all</u> $\lambda \in [0,1]$.

<u>In particular,</u>

(9.4.7) $\quad \Psi(\bar{u},1) = 0$

<u>has a solution in</u> Y_o .

Lemma 9.4 implies Thm. 9.5 via Lemma 9.3.

<u>Proof of Lemma 9.4:</u> That the transformation is of Leray - Schauder type, follows from Lemma 9.1. Consequently, the degree is well defined.

We now show that $\deg(\Psi(\bar{u},\lambda),Y_o,0)$ is independant of λ . Indeed, if $\Psi(\bar{u},\lambda) = 0$ for some $\bar{u} \in \bar{Y}_o$, then $|J(u)| > 0$ in $\bar{\Omega}$ by definition of Y_o . Consequently, $\bar{u} \in K(\lambda)$ by Lemma 9.3. Thms. 9.2 and 9.3 imply $K(\lambda) \subset Y_o$, and thus $\bar{u} \in Y_o$. Therefore, $\psi(\bar{u},\lambda) = 0$ cannot have a solution $\bar{u} \in \partial Y_o$, and the degree is independant of Ω .

It only remains to show that

$$(9.4.8) \qquad \deg(\Psi(\bar{u},0),Y_o,0) = 1 \ .$$

If we define $g(\varphi) = k(\varphi)$ for $\varphi \in \partial\Omega$, then

$$\Psi(\bar{u},0) = \Phi(\bar{u},1),$$

where the transformation Φ was defined in 9.3.
By Lemma 9.2 therefore

$$\deg(\Psi(\bar{u},0),Y,0) = 1 \ .$$

We note that $Y_o \subset Y$ (Y was defined in 9.3, too).
On the other hand, by the uniqueness theorem of Jäger - Kaul (cf. Thm. 5.1), any solution $\bar{u} \in Y$ of the equation $\Psi(\bar{u},0) = 0$ has to coincide with the one solution we know already, namely the conformal map k . Consequently, any solution in Y of $\Psi(\bar{u},0) = 0$ is actually contained in Y_o . Hence (9.4.8) follows from Lemma 9.2 and the excision property of the Leray - Schauder degree (cf. [D1], p. 67).

q.e.d.

Remark: Actually, one can dispense of the conformal map k in the preceding proof by using a more geometric variation of the boundary values, cf. [J1].

10. Harmonic coordinates. $C^{2,\alpha}$ - a - priori estimates for harmonic maps

10.1. Existence of harmonic coordinates. $C^{1,\alpha}$ - estimates

In this chapter, we want to prove existence and regularity of harmonic coordinates on surfaces, i.e. coordinates with harmonic coordinate functions. It will turn out that these harmonic coordinates possess optimal regularity properties. In particular, one can obtain C^α - bounds for the corresponding Christoffel symbols, in terms only of curvature bounds. This should be contrasted with the fact that for Riemannian normal coordinates, even L_∞ - bounds for the Christoffel symbols involve curvature derivatives (cf. [K1]).
Actually, in [JK1], there was displayed the following example of a twodimensional metric with Hölder continuous curvature, which itself is only Hölder continuous in normal coordinates, but not better:

$$ds^2 = dr^2 + G^2(r,\varphi)d\varphi^2$$

with

$$G^2(r,\varphi) = \begin{cases} r^2(1 + r^2 \sin^\alpha\varphi)^2 & \text{for } 0 \le \varphi \le \pi \quad (0 < \alpha < 1) \\ r^2 & \text{for } \pi \le \varphi \le 2\pi . \end{cases}$$

For this metric

$$K = -\frac{G_{rr}}{G} = \begin{cases} \dfrac{-6 \sin^\alpha\varphi}{1 + r^2 \sin^\alpha\varphi} & \text{for } 0 \le \varphi \le \pi \\ 0 & \text{for } \pi \le \varphi \le 2\pi . \end{cases}$$

The reason for this phenomenon is that the formula for K in normal coordinates does not involve any derivatives of G with respect to φ . A more detailed discussion of harmonic coordinates can be found in [JK1] and [JK2]. It should be mentioned that the optimal regularity properties of harmonic coordinates were noted by de Turck - Kazdan [dTK] and were made quantitively explicit in [JK1]. In higher dimensions, a disadvantage of harmonic coordinates is that (at least with the techniques presently available) one can prove their existence only on small balls, the size of these balls decreasing as the dimension increases. This is different in two dimensions, since here we can use the existence theorem 8.1 to obtain

Theorem 10.1: Suppose B is a bounded domain on a surface Σ , B being of the topological type of the disc, and having a boundary ∂B of class $C^{1,\alpha}$. Then there exists a canonically defined coordinate map h: B → D , D being the unit disc in the plane, which is harmonic, i.e. $\tau h = 0$.

Furthermore

(10.1.1) $$|h|_{C^{1,\alpha}} \le c_4 ,$$

c_4 depending on ω , κ ($-\omega^2 \le K \le \kappa^2$ on B), diam B , and the regularity of ∂B .

Remark: The regularity result will be improved to $C^{2,\alpha}$ in the next section.

Proof: If we take any map φ which maps ∂B onto ∂D proportionally to arclength, then we can extend φ to a harmonic diffeomorphism h_φ: B → D by Thm. 8.1 ($h_\varphi|\partial B = \varphi|\partial B$).

In local coordinates, since the image is flat

$$\frac{1}{\sqrt{\gamma}}\frac{\partial}{\partial x^\alpha}(\gamma^{\alpha\beta}\sqrt{\gamma}\frac{\partial}{\partial x^\beta}h^i) = 0 \qquad (i = 1,2) .$$

Choosing coordinates as in Lemma 2.5, e.g., on B, the coefficients of the Laplace - Beltrami operator are Hölder continuous, and (10.1.1) for h_φ then is a consequence of standard linear elliptic theory.
In order to get the harmonic coordinates canonically defined, we average over all maps φ (with the property described above) with a fixed orientation. In formulae, if $R(\vartheta)$ is the rotation of the plane by an angle of ϑ, then, choosing any such map φ,

$$h(x) := \frac{1}{2\pi}\int_{\vartheta=0}^{2\pi} R(-\vartheta)h_{R(\vartheta)\circ\varphi}(x)\,d\vartheta ,$$

is independant of the parametrization φ.
Clearly, the estimate (10.1.1) holds for h, since it holds uniformly for all $h_{R(\vartheta)\circ\varphi}$.

q.e.d.

Remarks: Other harmonic coordinates can be provided by conformal maps. The ones chosen here have the advantage that their boundary behaviour can be much better controlled.

10.2. $C^{2,\alpha}$ - estimates for harmonic coordinates

We now want to prove $C^{2,\alpha}$ - a - priori estimates for harmonic coordinate a result due to Jost - Karcher [JK1].

Theorem 10.2: Suppose $h: B_o \to D$ is a harmonic diffeomorphism of B_o onto its image with $J(h) \geq \delta^{-1}$ and $|dh| \leq c$ (Here, B_o is a bounded domain on a surface Σ, and D again is the unit disc). If ∂B_o and $h|\partial B_o$ are of class $C^{2,\alpha}$, then

(10.2.1) $$|h|_{C^{2,\alpha}(B_o)} \leq c_5 ,$$

$$c_5 = c_5(\omega,\kappa,\mathrm{diam}\, B_o,\delta,c,|h|\partial B_o|_{C^{2,\alpha}},|\partial B_o|_{C^{2,\alpha}})$$

Proof: We first need some calculations. h defines coordinates on B_o, and for the corresponding metric tensor (g_{ij}) we have

(10.2.2) $g^{ij} = \langle \text{grad } h^i , \text{grad } h^j \rangle$

and consequently

(10.2.3) $d_X g^{ij} = \langle D_X \text{ grad } h^i , \text{grad } h^j \rangle + \langle \text{grad } h^i , D_X \text{ grad } h^j \rangle$

and in particular

(10.2.4) $\langle D_{\text{grad } h^k} \text{ grad } h^i , \text{grad } h^j \rangle$

$= 1/2 (d_{\text{grad } h^k} g^{ij} - d_{\text{grad } h^i} g^{jk} + d_{\text{grad } h^j} g^{ik})$

Since h is harmonic,

(10.2.5) $D^2_{e^i, e^i} h^k = 0 \quad , \quad h = 1,2$

where e^i is an orthonormal frame.
Differentiating (10.2.5) w.r.t. e^j , we obtain

$$0 = \langle D^2_{e^j, e^i} \text{ grad } h^k , e^i \rangle$$
$$= \langle D^2_{e^i, e^i} \text{ grad } h^k , e^j \rangle + K \langle \text{grad } h^k , e^j \rangle ,$$

i.e.

(10.2.6) $\Delta \text{grad } h = K \cdot \text{grad } h$

From (10.2.3) and (10.2.6)

(10.2.7) $\Delta g^{ij} = 2Kg^{ij} + \langle D_{e^k} \text{ grad } h^i , e^l \rangle \langle e^l , D_{e^k} \text{ grad } h^j \rangle$

$= 2Kg^{ij} + g_{ms}\, g_{rt} \langle D_{\text{grad } h^m} \text{ grad } h^i , \text{grad } h^r \rangle \cdot$

$\langle \text{grad } h^s , D_{\text{grad } h^t} \text{ grad } h^j \rangle ,$

and therefore

(10.2.8) $|\Delta g^{ij}| \leq 2|K| \cdot |dh|^2 + \delta^2 |dh|^2 |d\tilde{g}|^2 ,$

using (10.2.4) (Here, $\tilde{g}$ denotes the matrix (g^{ij})).
If $k\colon D \to B(p,R) \subset \Sigma$ is conformal, then from (10.2.8)

$$(10.2.9)\qquad |\Delta(g^{ij}\circ k)| \leq 2|K||dh|^2\cdot\eta + \delta^2\eta|dh|^2|d\tilde{g}|^2 ,$$

if

$$(10.2.10)\qquad J(k) \geq \eta^{-1} > 0 .$$

We define $l: = \tilde{g}\circ k: D \to \mathbb{R}^3$
Putting in case $B(x_o, 2R_o) \subset D$

$$\mu: = \max_{x\in B(x_o, R_o)} (R_o - d(x,x_o))|dl(x)| ,$$

there exists $x_1 \in B(x_o R_o)$ with

$$(10.2.11)\qquad \mu = d\cdot|dl(x_1)| ,$$

where $d = R_o - d(x_1, x_o)$
and

$$(10.2.12)\qquad |dl(x_o)| \leq \frac{\mu}{R_o} .$$

As in the proof of Thm. 6.1, we obtain

$$\frac{\mu}{d} \leq \frac{1}{\pi d^2 \vartheta^2} \oint_{\partial B(x_1, d\vartheta)} |l(x) - l(x_1)||dx|$$

$$+ \frac{1}{2\pi} \int_{B(x_1, d\vartheta)} \frac{|\Delta l|}{d(x,x_1)} dx .$$

Using (10.2.9), the assumption $|dh| \leq c$, and assuming an upper bound for $|dk|$, i.e.

$$(10.2.13)\qquad |dk| \leq k_o ,$$

and defining

$$\delta = \delta(\vartheta): = \max_{x\in B(x_1, d\vartheta)} d(l(x), l(x_1)),$$

we obtain as in the proof of Thm. 6.1

$$(10.2.14)\qquad \mu \leq \frac{a}{2}\frac{\delta(\vartheta_o)}{\vartheta} + \frac{b}{2}\vartheta\mu^2 ,$$

and arguing as there,

$$\mu \le \frac{a\delta(\vartheta_o)}{\vartheta_o} , \tag{10.2.15}$$

since again $\delta(\vartheta_o)$ can be made arbitrarily small, choosing $\vartheta_o > 0$ sufficiently small, by (10.2.13) and $|dh| \le c$.
By Thms. 3.1, 6.1 and 7.1, we can always find a conformal map k satisfying (10.2.10) and (10.2.13) in discs interior to D .
Then (10.2.12), (10.2.15) and (10.2.10) imply

$$|d\tilde{g}| \le c_6 . \tag{10.2.16}$$

Going back to (10.2.8), we infer by a result from linear elliptic theory

$$|g^{ij}|_{C^{1,\alpha}} \le c_7 \tag{10.2.17}$$

(In order to apply this result from elliptic theory, we have to ensure that the C^α-norms of the coefficients of the Laplace -Beltrami operator Δ on B_o are bounded. This is no problem, since we can use the coordinates on B_o given by h , for which we already proved (10.2.16), i.e. a C^1 - bound on the coefficients of the inverse metric tensor (g^{ij}), and for which $g := \det(g_{ij})$ is controlled by the assumption $J(h) \ge \delta^{-1}$) .
Since $g^{ij} = \langle \operatorname{grad} h^i , \operatorname{grad} h^j \rangle$, (10.2.17) implies (10.2.1).

q.e.d.

10.3. Bounds on the Christoffel symbols. Conformal coordinates

Corollary 10.1: Suppose B, a bounded topological disc on a surface Σ , has a boundary ∂B of class $C^{2,\alpha}$. Then there exists a harmonic coordinate map h: B → D with

$$|h|_{C^{2,\alpha}} \le c_7 , \tag{10.3.1}$$

$c_7 = c_7(\omega,\kappa,\operatorname{diam} B, |\partial B|_{C^{2,\alpha}})$.
In particular, for the Christoffel symbols in the corresponding coordinate representation,

$$|\Gamma^i_{jk}|_{C^\alpha} \le c_8 , \tag{10.3.2}$$

$c_8 = c_8(\omega, \kappa, \text{diam } B, |\partial B|_{C^{2,\alpha}})$

Proof: The result is a consequence of Thms. 10.1, 10.2, 9.1 and 9.2.
(10.3.2) follows, since the Christoffel symbols are given by $D^2 h$, cf. (2.6.3).

Corollary 10.2: Let $\tau: B \to S$ be the conformal representation constructed in Thm. 3.1. We assume that S is of class C^3. If $B_o \subset\subset B$, then

(10.3.3) $$|\tau|_{C^{2,\alpha}(B_o)} \leq c_9 ,$$

$c_9 = c_9(\omega, \kappa, \text{diam } S, \text{dist}(B_o, \partial B))(-\omega^2 \leq K \leq \kappa^2 \text{ on } S)$.
In particular, in isothermal coordinates, we have interior bounds on the Christoffel symbols Γ^i_{jk}, depending only on curvature bounds, diameter and boundary distance.

Proof: The assertion again is a consequence of Thms. 10.2 and 9.1.

10.4. Higher regularity of harmonic coordinates

Corollary 10.3: Suppose $h: B \to D$ is a harmonic coordinate map. If the curvature of B is of class C^k or $C^{k,\beta}$, then h is of class $C^{k+2,\alpha}$ or $C^{k+3,\beta}$, resp., and we have corresponding a-priori estimates.
If the curvature is of class C^∞ or real analytic, then so is h.
(At the boundary, these results hold provided ∂B is sufficiently regular).
In particular, we have such regularity results for conformal maps.

Proof: Higher regularity follows by differentiating (10.2.7).

q.e.d.

10.5. $C^{2,\alpha}$-estimates for harmonic maps

Using harmonic coordinates, one can now improve Thms. 9.1 and 9.2 by a full order of differentiation, i.e. we can pass from $C^{1,\alpha}$ to $C^{2,\alpha}$-estimates.

Theorem 10.3: Under the assumptions of Thm. 9.1,

(10.5.1) $$|u|_{C^{2,\alpha}(\Omega_0)} \le c_{15} ,$$

where c_{15} depends on the same quantities as c_{10} and on nothing else.

Proof: For any $x_0 \in \Omega_0$, we can introduce harmonic coordinates on $B(x_0,R)$, $R: = \min(i(\Omega), d(x_0,\partial\Omega))$, by Cor. 10.1. Likewise, we can introduce harmonic coordinates on $B(p,M)$. If we write the equations for u in these coordinates, i.e.

(10.5.2) $$\frac{1}{\sqrt{\gamma}} \frac{\partial}{\partial x^\alpha} (\sqrt{\gamma}\, \gamma^{\alpha\beta} \frac{\partial u^i}{\partial x^\beta}) + \gamma^{\alpha\beta}\, \Gamma^i_{jk} \frac{\partial u^j}{\partial x^\alpha} \frac{\partial u^k}{\partial x^\beta} = 0 \quad (i = 1,2),$$

we see immediately from linear elliptic regularity theory, that (9.1.1) and (10.3.1) imply (10.5.1).

q.e.d.

In a similar way, we derive

Theorem 10.4: If under the assumptions of Thm. 9.2 $g: = u|\partial\Omega \in C^{2,\alpha}$ and $\partial\Omega \in C^{2,\alpha}$, then

(10.5.3) $$|u|_{C^{2,\alpha}(\Omega)} \le c_{16} ,$$

where c_{16} depends on the same quantities as c_{14} and on $|g|_{C^{2,\alpha}}$ and $|\partial\Omega|_{C^{2,\alpha}}$ and on nothing else.

10.6. Higher regularity of harmonic maps

Using Cor. 10.3, we finally obtain

Theorem 10.5: If under the assumptions of Thm. 9.1, the curvature of domain and image is of class C^k or $C^{k,\beta}$, then u is of class $C^{k+2,\alpha}$ or $C^{k+3,\beta}$, resp.
Corresponding a-priori estimates depend on the quantities mentioned in Thm. 9.1 and on the C^k or $C^{k,\beta}$-norms of the curvatures.
The same assertions hold at the boundary under the assumptions of Thm. 9.2, provided g and $\partial\Omega$ are sufficiently regular.
If the data are C^∞ or real analytic, then so is u.

11. The existence of harmonic diffeomorphisms between surfaces

11.1. Harmonic diffeomorphisms between closed surfaces (Theorem 11.1)

The main result of this chapter is

Theorem 11.1: Suppose that Σ_1 and Σ_2 are compact surfaces without boundary, and that $\varphi: \Sigma_1 \to \Sigma_2$ is a diffeomorphism. Then there exists a harmonic diffeomorphism $u: \Sigma_1 \to \Sigma_2$ homotopic to φ. Furthermore, u is of least energy among all diffeomorphisms homotopic to φ.

Thm. 11.1 was proved by Jost - Schoen [JS], but it was first claimed by Shibata [Sh] in 1963. His proof contained several mistakes, however, and was therefore rejected.

H. Sealey then carefully examined Shibata's paper in his thesis [Se] and was able to correct some (but not all) of the mistakes. The proof of [JS], however, proceeds along completely different lines than the Shibata - Sealey approach and depends in an essential way on Thm. 8.1. Thms. 11.1 and 5.3 immediately imply the following corollary, proved by Schoen - Yau [SY1] and Sampson [Sa].

Corollary 11.1: If under the assumptions of Thm. 11.1, Σ_2 has non-positive curvature, then every harmonic map homotopic to a diffeomorphism is itself diffeomorphic.

Furthermore, we have

Corollary 11.2: Suppose that Σ_1 and Σ_2 are compact surfaces without boundary, and that $\psi: \Sigma_1 \to \Sigma_2$ is a covering map, i.e. a local diffeomorphism. Then there exists a harmonic covering map $u: \Sigma_1 \to \Sigma_2$, homotopic to ψ.

Proof of Corollary 11.2: We can pull back the metric ds^2 of Σ_2 via ψ to obtain a surface Σ_2', diffeomorphic to Σ_1 and with metric ψ^*ds^2. Then $\psi: \Sigma_2' \to \Sigma_2$ is a local isometry. By Thm. 11.1, there is a harmonic diffeomorphism $u': \Sigma_1 \to \Sigma_2'$, homotopic to the identity. $u := \psi \circ u$ then is the desired harmonic covering map.

11.2 Proof of Theorem 11.1

Proof of Theorem 11.1: (following [JS]). If Σ_1 and Σ_2 are homeomorphic to S^2, then we can find a conformal (and hence harmonic) diffeomorphism homotopic to ψ, since any two metrics on S^2 are conformally equivalent by the Theorem of Riemann - Roch. The case where Σ_1 and Σ_2 are homeomorphic to the real projective space is similarly handled by passing to two - sheeted coverings. Thus we can assume w.l.o.g. that $\pi_2(\Sigma_i) = 0$ $(i = 1,2)$.

Let D_K be the class of all diffeomorphisms from Σ_1 onto Σ_2 homotopic to ϕ the energy of which is bounded by K , where K is chosen sufficiently large that D_K is not empty. Since $\pi_2(\Sigma_i) = 0$ $(i = 1,2)$, we can lift every map in D_K to a map between the universal covers of the Σ_i , which are discs, topologically, and apply the argument of section 3.3 to conclude that D_K is equicontinuous.

Let $\overline{D}_K$ be the closure of D_K with respect to weak H_2^1 convergence. By the lower semicontinuity of the energy w.r.t. weak H_2^1 convergence, all elements in $\overline{D}_K$ have energy bounded by K . Let $(t_n)_{n\in N}$ be a sequence in $\overline{D}_K$ with

$$E(t_n) \to \inf_{t\in D_K} E(t) \qquad \text{for } n \to \infty$$

w.l.o.g. (or more precisely, by passing to a subsequence) we can assume that t_n converges weakly in the H_2^1 topology and uniformly in the C^0 topology to a map u_0 in the prescribed homotopy class, using the weak compactness of H_2^1 and the Arzela - Ascoli Theorem. By the lower semicontinuity of the energy again,

$$E(u_0) = \inf_{t\in D_K} E(t)$$

Also, we can find a sequence of diffeomorphisms $u_n \in D_K$ which converge weakly in H_2^1 and uniformly to u_0 .

We want to show that u_0 is a harmonic diffeomorphism. We consider an arbitrary point $x_0 \in \Sigma_1$ and define

$$B_\sigma := \overset{0}{B}(u_0(x_0),\sigma)$$

i.e. the open disc in Σ_2 centered at $u_0(x_0)$ with radius σ .

We restrict ourselves in the sequel to values of σ which are smaller than the injectivity radius of Σ_2 and smaller than $\pi/2\kappa$, where κ^2 again is an upper bound for the curvature of Σ_2 . We define

$$\Omega_0 := u_0^{-1}(B_\sigma)$$

$$\Omega_n := u_n^{-1}(B_\sigma) \qquad (n \in N)$$

W.l.o.g., we can assume $x_o \in \Omega_n$ for all n, since the u_n converge uniformly to u_o. Let D be the unit disc in the complex plane and

$$F_n : D \to \bar{\Omega}_n$$

be a conformal mapping which maps O to x_o.
The proof of the existence of F_n is the same as that of Thm. 3.1, since instead of fixing three boundary points, we can fix an interior point (and a tangent direction at this point, but that is not necessary for the proof) in order to guarantee the equicontinuity of a minimizing sequence as in 3.3.
Since $\Gamma_n := \partial\Omega_n$ is a Jordan curve of class C^1 (because u_n is a diffeomorphism), F_n is a homeomorphism of D onto $\bar{\Omega}_n$, and therefore $u_n \circ F_n$ maps ∂D homeomorphically onto ∂B_σ. By Thm. 8.1 and Cor 8.1, there exists a unique harmonic mapping $v_n : D \to B_\sigma$ which assumes the boundary values prescribed by $u_n \circ F_n$, and v_n minimizes energy in its homotopy class and is a diffeomorphism.
In particular,

$$\text{(11.2.1)} \qquad E_D(v_n) \leq E_D(u_n \circ F_n) = E_{\Omega_n}(u_n) \leq K$$

by Lemma 6.1. ($E_S(f)$ is the energy of the mapping f over the set S).
Since the u_n converge uniformly to u_o, we can assume that $u_n \circ F_n(O)$ stays in an arbitrarily small neighborhood of $u_o(x_o)$. Therefore, we can again apply the argument of section 3.3 to show that the maps $u_n \circ F_n$ are equicontinuous on D. In particular, the boundary values of v_n, namely $u_n \circ F_n | \partial D$, are equicontinuous. By Thm. 10.3 and Cor. 6.2, we can therefore assume that the v_n converge uniformly on D to a map v_o which is harmonic in the interior of D. Using Cor. 7.1, we see furthermore, that v_o is a diffeomorphism in the interior of D.
We define now

$$\tilde{u}_n = \begin{cases} v_n \circ F_n^{-1} & \text{in } \Omega_n \\ u_n & \text{in } \Sigma_1 \setminus \Omega_n \end{cases}$$

Clearly, $\tilde{u}_n$ is a Lipschitz map and lies in H_2^1 and $E(\tilde{u}_n) \leq K$. We can also assume w.l.o.g. (by approximation) that the u_n are of class $C^{1,\alpha}$. Then, for each n, the functional determinant of $\tilde{u}_n$ is defined and bounded from below on $\Sigma_1 \setminus \Gamma_n$ by Cor. 7.2.
If u_n and $v_n \circ F_n^{-1}$ do not coincide on Ω_n, then $\tilde{u}_n$ is not necessarily smooth on Γ_n, but in this case $E(\tilde{u}_n)$ is strictly less than K by uniqueness of the energy minimizing map, and an approximation argument shows that $\tilde{u}_n$ can be approximated in the H_2^1 norm by diffeomorphisms with energy bounded by K, and $\tilde{u}_n$ lies therefore in $\overline{D}_K$ in any case.
Using Lemma 3.1, we can assume again w.l.o.g. that the $\tilde{u}_n$ converge on Σ_1 weakly in H_2^1 and uniformly to a map $\tilde{u}_0 \in \overline{D}_K$ and that the F_n converge uniformly on compact subsets to a conformal map F. Since $E_D(F_n) = \text{Area}(\Omega_n) \leq \text{Area}(\Sigma_1)$, F maps $\overset{\circ}{D}$ diffeomorphically onto some open set $\Omega \subset \Sigma_1$, and 0 is mapped to x_0. F is not necessarily smooth on ∂D, but that does not affect the following arguments.
$u_0 \circ F$ is the uniform limit of $u_n \circ F_n$ and thus extends continuously to D. Since $u_n \circ F_n$ and v_n coincide on ∂D, it follows that also $u_0 \circ F$ and v_0 coincide there, and since v_0 is harmonic and therefore energy minimizing (by Theorem 5.1) in its homotopy class,

$$E_D(v_0) \leq E_D(u_0 \circ F)$$

Since conformal maps preserve energy by Lemma 6.1, this implies

(11.2.2) $$E_\Omega(\tilde{u}_0) \leq E_\Omega(u_0)$$

We now want to show that

(11.2.3) $$E_{\Sigma_1 \setminus \Omega}(\tilde{u}_0) = E_{\Sigma_1 \setminus \Omega}(u_0)$$

For this, it is sufficient to show that u_0 and $\tilde{u}_0$ coincide almost everywhere outside Ω. We claim that

(11.2.4) $$\Sigma_1 \setminus \Omega \subset u_0^{-1}(\Sigma_2 \setminus B_\sigma)$$

We define

$$\rho_n(x) := d(u_n(x), u_o(x_o))$$

$$\rho_o(x) := d(u_o(x), u_o(x_o))$$

for $x \in \Sigma_1$. Let $x \in \Sigma_1 \setminus \Omega$. If

$$\rho_o(x) = \lim_{n\to\infty} \rho_n(x) \geq \sigma \text{ , then}$$

$$x \in u_o^{-1}(\Sigma_2 \setminus B_\sigma)$$

Since the $\rho_n \circ u_n \circ F_n$ are equicontinuous and equal to σ on ∂D , $\rho_o(x) <$ implies that

$$d(F_o^{-1}(x), \partial D) \geq \delta > 0$$

for sufficiently large n .
Since on the other hand, the F_n converge uniformly to F on compact sub sets of D , this would imply $x \in F(D) = \Omega$ which contradicts the assumption $x \in \Sigma_1 \sim \Omega$. This proves (11.2.4)
We also have

$$u_o^{-1}(\Sigma_2 \setminus B_\sigma) = u_o^{-1}(\partial B_\sigma) \cup u_o^{-1}(\Sigma_2 \setminus \overline{B}_\sigma) ,$$

and since the sets $u_o^{-1}(\partial B_\sigma)$ cover a neighborhood of x_o and are disjoint, we can assume w.l.o.g. that the twodimensional measure of u_o^{-1} (∂B_σ) vanishes for one chosen σ . If

$$x \in u_o^{-1}(\Sigma_2 \setminus \overline{B}_\sigma),$$

then

$$\lim_{n\to\infty} \rho_n(x) = \rho_o(x) > \sigma .$$

and because of the equicontinuity of the functions ρ_n , there exists an open neighborhood U of x such that $\rho_n | U > \sigma$ for sufficiently large n . This implies

$$\tilde{u}_o = \lim_{n\to\infty} \tilde{u}_n = \lim_{n\to\infty} u_n = u_o \quad \text{on } U .$$

Therefore $u_o = \tilde{u}_o$ almost everywhere on $u_o^{-1}(\Sigma_2 \setminus B_\sigma)$, and (11.2.3) now follows from (11.2.4). By the choice of u_o , we have on the other han

$$E_{\Sigma_1}(u_o) \leq E_{\Sigma_1}(\tilde{u}_o)$$

Thus, we conclude from (11.2.2) and (11.2.3) that

$$E_\Omega(\tilde{u}_o) = E_\Omega(u_o)$$

and consequently

$$E_D(v_o) = E_D(u_o \circ F) \ .$$

Since v_o and $u_o \circ F$ coincide on ∂D , we conclude from the uniqueness of energy minimizing maps (Thms. 4.1 and 5.1) that v_o and $u_o \circ F$ coincide on D . Therefore $u_o \circ F$ and consequently also u_o is a harmonic diffeomorphism, the latter in Ω , which is a neighborhood of an arbitrarily chosen point $x_o \in \Sigma_1$. This finishes the proof of Theorem 11.1.

11.3. Extension of Theorem 8.1

With the same method, we can also improve Thm. 8.1

Theorem 11.2: Let $\Omega \subset \Sigma_1$ be a twodimensional domain with nonempty boundary $\partial\Omega$ consisting of Lipschitz curves, and let $\psi: \bar{\Omega} \to \Sigma_2$ be a homeomorphism of $\bar{\Omega}$ onto its image $\psi(\bar{\Omega})$, and suppose that the curves $\psi(\partial\Omega)$ are of Lipschitz class and convex with respect to $\psi(\Omega)$. Then there exists a harmonic diffeomorphism $u: \Omega \to \psi(\Omega)$ which is homotopic to ψ and satisfies $u = \psi$ on $\partial\Omega$. Moreover, u is of least energy among all diffeomorphisms homotopic to ψ and assuming the same boundary values.

This result is taken from [JS]. The case of non - positive image curvature was solved in [SY1].

Proof: We assume first that $\partial\Omega$ and $\psi(\partial\Omega)$ are of class $C^{2+\alpha}$ and that ψ gives rise to a diffeomorphism between $\partial\Omega$ and $\psi(\partial\Omega)$ and that $\psi(\partial\Omega)$ is strictly convex with respect to $\psi(\Omega)$.

In this case, the proof proceeds along the lines of the proof of Theorem 11.1 with an obvious change of the replacement argument at boundary points involving Thm. 7.2. The general case now follows by approximation arguments as in 8.2.

11.4. Remarks about the situation in higher dimensions

As shown in 7.4, in higher dimensions one cannot expect an analogue of

Thm. 11.1 or even of Cor. 11.1 to hold. Since in the example of 7.4, the image was flat, however, the case where the image has strictly negative curvature still remains open.
On the other hand, in the context of Kähler manifolds, i.e. Riemannian manifolds, carrying a complex structure compatible with its Riemannian structure, it was possible to show in some cases, that the (unique) harmonic map in a given homotopy class is necessarily a diffeomorphism, cf. [Si] and [JY]. The image manifolds considered in these papers seem to be rather special, however, and thus the question of the existence of a harmonic diffeomorphism in the Kähler setting has a satisfactory answer only in complex dimension 1, since one-dimensional compact Kähler manifolds are nothing but compact orientable surfaces with a suitable metric and corresponding conformal structure.

12. Applications of harmonic maps between surfaces

12.1. Holomorphicity of certain harmonic maps and an analytic proof of Kneser's Theorem

Suppose that Σ_1 and Σ_2 are closed orientable surfaces, $\chi(\Sigma)$ denotes the Euler characteristic of a surface Σ, and $d(\varphi)$ is the degree of a map φ.
In [EW1], Eells and Wood obtained the following result:

Theorem 12.1: Suppose $h: \Sigma_1 \to \Sigma_2$ is harmonic with respect to metrics γ and g on Σ_1 and Σ_2, resp. If

$$\chi(\Sigma_1) + |d(h)|\,|\chi(\Sigma_2)| > 0 ,$$

then h is holomorphic or antiholomorphic relative to the complex structures determined by γ and g.
Thm. 12.1 enabled Eells and Wood to give an analytical proof of the following topological result of H. Kneser [Kn2]

Theorem 12.2: Suppose again that Σ_1 and Σ_2 are closed orientable surfaces, and furthermore $\chi(\Sigma_2) < 0$. Then for any continuous map $\varphi: \Sigma_1 \to \Sigma_2$

$$|d(\varphi)|\chi(\Sigma_2) \geq \chi(\Sigma_1) . \tag{12.1.1}$$

Proof of Theorem 12.2: We introduce some metrics γ and g on Σ_1 and

Σ_2 , resp., and find a harmonic map h homotopic to φ by Thm. 4.2. By Thm. 12.1, h is (anti) holomorphic in case $|d(\varphi)|\chi(\Sigma_2) < \chi(\Sigma_1)$. This, however, is in contradiction to the Riemann - Hurwitz formula, which says $|d(h)|\chi(\Sigma_2) = \chi(\Sigma_1) + r$, $r \geq 0$ for an (anti) holomorphic map h . Therefore, (12.1.1) must hold.

q.e.d.

Before proving Thm. 12.1, we note two other interesting consequences

Corollary 12.1: If Σ_1 is diffeomorphic to S^2 , then any harmonic map $h: \Sigma_1 \to \Sigma_2$ is (anti)holomorphic (and therefore constant, if $\chi(\Sigma_2) \leq 0$).

This is due to Wood [W1] and Lemaire [L1].

Corollary 12.2: If Σ_1 is diffeomorphic to the torus, and Σ_2 to S^2 , then there is no harmonic map $h: \Sigma_1 \to \Sigma_2$ with $d(h) = \pm 1$, for any metrics on Σ_1 and Σ_2 .

Cor. 12.2, due to Eells - Wood, follows from Thm. 12.1, since any holomorphic map of degree 1 is a covering map.

12.2. Proof of Theorem 12.1

In this section, we want to prove Thm. 12.1.
We shall make use of some computations of Schoen and Yau [SY1] in the sequel.
It is convenient to use the complex notation.
If $\rho^2(z)dzd\bar{z}$ and $\sigma^2(h)dhd\bar{h}$ are the metrics w.r.t. to conformal coordinate charts (for the existence, cf. Thm. 3.1) on Σ_1 and Σ_2 , resp., then h as a harmonic map satisfies

(12.2.1) $$h_{z\bar{z}} + \frac{2\sigma_h}{\sigma} h_z h_{\bar{z}} = 0 \text{ , cf. (1.3.4)}$$

Lemma 12.1: At points, where ∂h or $\bar{\partial} h$, resp., is non zero

(12.2.2) $$\Delta \log |\partial h|^2 = K_1 - K_2(|\partial h|^2 - |\bar{\partial} h|^2)$$

(12.2.3) $$\Delta \log |\bar{\partial} h|^2 = K_1 + K_2(|\partial h|^2 - |\bar{\partial} h|^2) ,$$

where K_i denotes the Gauss curvature of Σ_i , and

$$|\partial h|^2 = \frac{\sigma^2}{\rho^2} h_z \cdot \bar{h}_{\bar{z}} \quad , \quad |\bar{\partial} h|^2 = \frac{\sigma^2}{\rho^2} \bar{h}_z h_{\bar{z}}$$

<u>Proof:</u> For any positive smooth function f on Σ_1,

(12.2.4) $$\Delta \log f = \frac{1}{f} \Delta f - \frac{1}{f^2} \cdot \frac{1}{\rho^2} f_z f_{\bar{z}} \ .$$

Furthermore,

(12.2.5) $$\Delta \log \frac{1}{\rho^2} = K_1 \ .$$

In order to abbreviate the following calculations, we define D as the covariant derivative in the bundle $h^{-1}\, T\Sigma_2$, e.g.

$$D_{\frac{\partial}{\partial z}} h_z = h_{zz} + \frac{2\sigma_h}{\sigma} h_z h_z$$

(12.2.1) then is expressed as

(12.2.6) $$D_{\frac{\partial}{\partial z}} h_{\bar{z}} = 0$$

Since

$$\sigma^2 h_z \bar{h}_{\bar{z}} = \langle h_z , \bar{h}_{\bar{z}} \rangle_{h^{-1}T\Sigma_2} \ ,$$

(12.2.7) $$\Delta \sigma^2 h_z \bar{h}_z = \frac{1}{\rho^2} \frac{\partial}{\partial \bar{z}} \langle D_{\frac{\partial}{\partial z}} h_z , \bar{h}_{\bar{z}} \rangle, \text{ using (12.2.6)}$$

$$= \frac{1}{\rho^2} \langle D_{\frac{\partial}{\partial \bar{z}}} D_{\frac{\partial}{\partial z}} h_z , \bar{h}_{\bar{z}} \rangle + \frac{1}{\rho^2} \langle D_{\frac{\partial}{\partial z}} h_z , D_{\frac{\partial}{\partial \bar{z}}} \bar{h}_{\bar{z}} \rangle$$

$$= \frac{1}{\rho^2} R(h_*(\tfrac{\partial}{\partial \bar{z}}), h_*(\tfrac{\partial}{\partial z}), h_z , \bar{h}_{\bar{z}}) +$$

$$+ \frac{1}{\rho^2} \langle D_{\frac{\partial}{\partial z}} h_z , D_{\frac{\partial}{\partial \bar{z}}} \bar{h}_{\bar{z}} \rangle,$$

where R denotes the curvature tensor of Σ_2

$$= -K_2 |\partial h|^2 (J(h)) +$$

$$+ \frac{1}{\rho^2} \langle D_{\frac{\partial}{\partial z}} h_z , D_{\frac{\partial}{\partial \bar{z}}} \bar{h}_{\bar{z}} \rangle,$$

where $J(h) = |\partial h|^2 - |\bar{\partial} h|^2$ is the Jacobian of h .

Moreover,

(12.2.8) $$\frac{1}{\rho^2}\frac{\partial}{\partial z}\langle h_z , h_{\bar{z}}\rangle \cdot \frac{\partial}{\partial \bar{z}}\langle h_z , h_{\bar{z}}\rangle$$

$$= \frac{1}{\rho^2}\langle h_z , h_{\bar{z}}\rangle \quad \langle D_{\frac{\partial}{\partial z}} h_z , D_{\frac{\partial}{\partial \bar{z}}} h_{\bar{z}} \rangle ,$$

using again (12.2.6) and the fact that the complex dimension of Σ_2 is 1 .

(12.2.2) now follows from (12.2.4), (12.2.5), (12.2.7), and (12.2.8), and (12.2.3) can either be calculated in the same way or directly deduced from (12.2.2), since $|\bar{\partial} h|^2 = |\partial \bar{h}|^2$ and complex conjugation on the image can be considered as a change of orientation.

q.e.d.

Lemma 12.2: If $h_z(z_o) = 0$, then

(12.2.8) $$|\partial h|^2 = \zeta \cdot |k|^2 \quad \text{near } z = z_o ,$$

where ζ is a nonvanishing C^2 function, and k is holomorphic. A corresponding result holds for $h_{\bar{z}}$.

Proof: By (12.2.1), f: = h_z satisfies

$$|f_{\bar{z}}| \le c|f| .$$

Therefore, we can apply the similarity principle of Bers and Vekua (cf. [B2] or [Hz1]), to obtain the representation (12.2.8) with Hölder continuous ζ . An inspection of the proof of the similarity principle shows that in our case $\zeta \in C^2$ (cf. [Hz1], p. 210).

q.e.d.

Proof of Theorem 12.1: Lemma 12.2 shows that the zeroes z_i of $|\partial h|^2$ are isolated, unless $\partial h \equiv 0$, and that near each z_i ,

$$|\partial h|^2 = a_i |z - z_i|^{n_i} + o(|z - z_i|^{n_i}) ,$$

for some $a_i > 0$ and some $n_i \in \mathbb{N}$.

By Lemma 12.1 and the residue formula, unless $\partial h \equiv 0$

$$(12.2.9) \qquad \int_{\Sigma_1} K_1 - \int_{\Sigma_2} K_2(|\partial h|^2 - |\bar{\partial} h|^2) = -\Sigma n_i$$

Similarly, if $\bar{\partial} h \not\equiv 0$,

$$(12.2.10) \qquad \int_{\Sigma_1} K_1 + \int_{\Sigma_2} K_2(|\partial h|^2 - |\bar{\partial} h|^2) = -\Sigma m_i$$

where $m_i \in \mathbb{N}$ are now the orders of the zeros of $|\bar{\partial} h|^2$.

Thus, since $|\partial h|^2 - |\bar{\partial} h|^2$ is the Jacobian of h ,

$$\chi(\Sigma_1) - d(h)\,\chi(\Sigma_2) \le 0 \quad , \quad \text{unless } \partial h \equiv 0$$

and

$$\chi(\Sigma_1) + d(h)\chi(\Sigma_2) \le 0 \quad , \quad \text{unless } \bar{\partial} h \equiv 0 ,$$

and Thm. 12.1 follows.

12.3. Contractability of Teichmüller space and the diffeomorphism group

In the following sections, we want to give some applications of harmonic maps between surfaces to Teichmüller theory. The arguments in the sequel are due to Earle - Eells [EE] and Tromba [Tr].

We first recall the setting of Teichmüller theory.

Let Σ be a compact oriented smooth twodimensional manifold without boundary of genus at least 2 (The case of the sphere and the torus is classical and much easier to handle). Note that we deviate here from our previous notation, where Σ denoted a surface endowed with a Riemannian metric. Such a pair will be denoted by (Σ, g) in this section. $D(\Sigma)$ is the diffeomorphism group of Σ , while $D_0(\Sigma)$ is the subgroup of $D(\Sigma)$ consisting of those diffeomorphisms which are homotopic to the identity map id of Σ .

$M(\Sigma)$ is the space of smooth complex structures on Σ (compatible with the orientation). For a given complex structure we can look at the metrics compatible with these structures, and the uniformization theore easily shows that each element of $M(\Sigma)$ corresponds to precisely one metric of constant curvature -1 . Thus, we shall think of an element of $M(\Sigma)$ as a metric g on Σ of curvature -1 . We have a natural action

$$M(\Sigma) \times D(\Sigma) \to M(\Sigma)$$

defined by

$$(g,v) \to (v^*g) \ ,$$

where v^*g denotes the pull back of the metric g via v , i.e. if $x \in \Sigma$, then $v^*g(x) = g(v\ (x))$.
Then the Teichmüller space $T(\Sigma)$ is defined by

$$T(\Sigma) = M(\Sigma)/D_o(\Sigma).$$

$D(\Sigma)$ carries the C^∞ topology of uniform convergence of derivatives of all orders, $M(\Sigma)$ also carries its C^∞ topology, and $T(\Sigma)$ inherits the quotient topology.
The main result is Teichmüller's Theorem, saying that $T(\Sigma)$ is homeomorphic to $\mathbb{R}^{6g-6}$, where g is the genus of Σ . For an account of the theory, we refer to [Tm], [A1], [A2], [AB], [B1], [EE], and [FT].

The following result is due to Earle and Eells [EE].

Theorem 12.3: The bundle given by the quotient map $p: M(\Sigma) \to T(\Sigma)$ is trivial, i.e. there exists a homeomorphism F with the property that

$$\begin{array}{ccc} M(\Sigma) & \xrightarrow{F} & T(\Sigma) \times D_o(\Sigma) \\ & {\scriptstyle p}\searrow \quad \swarrow{\scriptstyle \pi} & \\ & T(\Sigma) & \end{array}$$

commutes, where π is the projection onto the first factor.

Proof: We fix some conformal structure, i.e. a metric γ of curvature -1 . Then for any other metric g on Σ of curvature -1 , there exists a harmonic diffeomorphism

$$u(\gamma,g) : (\Sigma,\gamma) \to (\Sigma,g)$$

by Theorem 11.1. $u(\gamma,g)$ is unique by Theorem 5.3. By uniqueness and the a-priori estimates of chapter 6, it follows that $u(\gamma,g)$ depends

continuously on g. The lower bounds for the functional determinant of harmonic diffeomorphisms of Thm. 7.1 then show that also $u(\gamma,g)^{-1}$ depends continuously on g.
(Continuous dependance in the present context seems to have been first noticed by Sampson).
Since by construction, $v: (\Sigma, v^*g) \to (\Sigma, g)$ $(v \in D_o(\Sigma))$, is an isometry, also $v \circ u(\gamma, v^*g)$ is harmonic between (Σ,γ) and (Σ,g), and homotopic to $u(\gamma,g)$. By Thm. 5.3,

$$(12.3.1) \qquad v \circ u(\gamma, v^*g) = u(\gamma, g) .$$

We now define F by

$$F(g) = (p(g), u(\gamma,g)^{-1}) .$$

As was noted above, F is continuous. Furthermore, from (12.3.1) and the definitions of p

$$(12.3.2) \qquad F(v^*g) = (p(v^*g), u(\gamma, v^*g)^{-1}) = (p(g), u(\gamma,g)^{-1} \circ v)$$

for any $v \in D_o$.
If $F(g) = F(g')$, then $p(g) = p(g')$ and therefore $g' = v^*g$ for some $v \in D_o(\Sigma)$ by definition of p, and $F(g') = (p(g), u(\gamma,g)^{-1} \circ v)$. Since $F(g') = F(g)$, thus $v = id$ and $g = g'$. This shows that F is injective. F is surjective, since

$$\begin{aligned}(p(g), v) &= (p(g), u(\gamma,g)^{-1} \circ u(\gamma,g) \circ v) \\ &= (p(g), u(\gamma, v^*u(\gamma,g)^*g)^{-1}) \text{ by } (12.3.1) \\ &= F(v^*u(\gamma,g)^*g) .\end{aligned}$$

Therefore, F is a homeomorphism, and the theorem follows.

q.e.d.

A local section $\sigma: T(\Sigma) \to M(\Sigma)$ is given by

$$\sigma(t) = u(\gamma,g)^*g ,$$

where we can take any g with $t = p(g)$, since $\sigma(t)$ is independant of this choice by (12.3.1) again. We now use the fact that $M(\Sigma)$ is contractible, since $M(\Sigma) = \mathcal{M}(\Sigma)/P(\Sigma)$, where $\mathcal{M}(\Sigma)$ is the space of Riemannian metrics on Σ , i.e. an open convex cone in the space of symmetric 2 - tensors on Σ , and $P(\Sigma)$ is the space of positive C^∞ - functions on Σ and both $\mathcal{M}(\Sigma)$ and $P(\Sigma)$ are contractible (cf. [FT].)
This implies the following Corollary to Thm. 12.3.

Corollary 12.3: $T(\Sigma)$ and $D_o(\Sigma)$ are contractible.

12.4. Tromba's proof that Teichmüller space is a cell

We now want to sketch an argument of Tromba [Tr] which shows that $T(\Sigma)$ is a cell. Of course, this is part of Teichmüller's Theorem, but Tromba's argument provides an easy proof by using harmonic maps. We continue to use the notations of the previous section.
The following theorem seems to have been known more or less for a long time. The proof we shall give here is taken from Sacks-Uhlenbeck [SkU]. Before stating the theorem, however, we formulate a lemma.

Lemma 12.3: Suppose $u(\gamma,g): (\Sigma,\gamma) \to (\Sigma,g)$ is harmonic. Then $u(\gamma,g)$ is also a critical point of the energy functional

$$E(u,\gamma,g) = 1/2 \int \gamma^{\alpha\beta}(x) g_{ij}(u) D_\alpha u^i D_\beta u^j \sqrt{\gamma}\, dx$$

with respect to all variations of the metric γ via a family $v(t)^*\gamma$ where $v(t)$ is a smooth family of diffeomorphisms of Σ with $v(0) = id$.

Proof: The reason why Lemma 12.3 holds is that the change of the metric by a diffeomorphism has the same effect as changing u by composition with the inverse diffeomorphism.
In formulae

$$\int \gamma^{\alpha\beta}(v(x)) g_{ij}(u(x)) D_\alpha u^i(x) D_\beta u^j(x) \sqrt{\gamma(v(x))}\, dx =$$

$$= \int \gamma^{\alpha\beta}(y) g_{ij}(u(v^{-1}(y))) D_\alpha u(v^{-1}(y)) D_\beta u(v^{-1}(y)) \sqrt{\gamma(y)}\, dy .$$

Since $u \circ v^{-1}(t)$ provides a variation of u , the lemma follows.

q.e.d.

Furthermore, we note that $E(u,\sigma\gamma,g) = E(u,\gamma,g)$ for any positive function σ on Σ , by Lemma 6.1. Therefore, if u is a critical point of E with respect to the metric γ , it is also critical with respect to all conformally equivalent metrics $\sigma\gamma$. Therefore, taking also Lemma 12.3 into account, if we want to investigate the effect of a variation of the domain metric on E at a critical point $u(\gamma,g)$, it is enough to vary the corresponding element $p'(\gamma)$ in $T(\Sigma)$. Here, $p'(\gamma)$ is defined by taking the class of metrics which are conformally equivalent to γ , i.e. the corresponding element of $M(\Sigma)$ and then projecting this conformal structure onto $T(\Sigma)$ by the quotient map p of the previous section. We can now state

Theorem 12.4: Suppose $u(\gamma,g)$ is a critical point of $E(u,\gamma,g)$ both with respect to variations of u and the conformal structure of γ in $T(\Sigma)$. Then $u(\gamma,g)$ is a conformal branched covering.

Proof: We proved already that u is a critical point with respect to any smooth change $\gamma(t)$ of the metric γ . For any point $p \in \Sigma$, a neighborhood of p can be conformally represented by the unit disc by Thm. 3.1. This provides local coordinates, and using Lemma 6.1 again, we can assume w.l.o.g. that the metric in these local coordinates has the form

$$\gamma_{\alpha\beta}(z) = \delta_{\alpha\beta} \qquad \text{for } z \in D .$$

If we now change γ by a smooth variation $\gamma(t)$ with

$$\gamma_{\alpha\beta}(t,z) = \delta_{\alpha\beta} \qquad \text{for } z \text{ in a neighborhood of } \partial D$$

and

$$\gamma_{11}(t,z) = \gamma_{22}(t,z) = 1 \text{ for all } t,z \quad ,$$

then

$$\frac{d}{dt} E(u,\gamma(t),g) = \frac{d}{dt} \int_D \gamma_{\alpha\beta}(t,z) g_{ij} D_\alpha u^i D_\beta u^j \sqrt{\gamma(t,z)}\, dz =$$

$$= \frac{d}{dt} \int_D g_{ij}(u)(D_1u^i D_1u^j + D_2u^i D_2u^j + 2\gamma^{12}(t,z)$$

$$D_1u^i D_2u^j) \cdot \sqrt{1-(\gamma^{12}(t,z))^2}\, dz =$$

$$= 2 \int_D \frac{\partial}{\partial t} \gamma^{12}(t,z) g_{ij}(u)\, D_1 u^i\, D_2 u^j\, dz$$

Since $\frac{d}{dt} E(u,\gamma(t),g) = 0$ at $t = 0$, as shown above, and since $\frac{\partial}{\partial t} \gamma^{12}(t,z)$ can be chosen arbitrarily at $t = 0$, we infer

$$g_{ij}(u)\, D_1 u^i\, D_2 u^j = 0 \quad .$$

If we rotate the isothermal coordinates by an angle of $\pi/4$, the same argument yields that also

$$g_{ij}(u)(D_1 u^i\, D_1 u^j - D_2 u^i\, D_2 u^j) = 0 \; .$$

Therefore, the quadratic differential

(12.4.1) $$\phi = (g_{ij}(u)(D_1 u^i\, D_1 u^j - D_2 u^i\, D_2 u^j - 2i\, D_1 u^i\, D_2 u^j) dz^2 =$$

$$= (|u_x|^2 - |u_y|^2 - 2i \langle u_x, u_y \rangle) dz^2$$

$(z = x + iy)$
vanishes identically, and the theorem follows.

q.e.d.

We now prove the following result of Schoen - Yau [SY2]:

Theorem 12.5: Suppose g is a fixed metric with nonpositive curvature on Σ, and define a function on $T(\Sigma)$ via

$$p \to E(u(\gamma_p, g)) \; ,$$

where γ_p is a metric representing p, and $u(\gamma_p, g)$ is the corresponding harmonic map from (Σ, γ_p) to (Σ, g), homotopic to the identity of Σ. Then E is a proper function on $T(\Sigma)$, and we can find a $p \in T(\Sigma)$ for which $u(\gamma_p, g)$ minimizes E with respect to both u and p.

Proof: Using Theorem 5.3 and the a - priori estimates, we see that E is a continuous function of p.

We now show that if γ is a metric of constant curvature -1 and (Σ,γ) contains a closed geodesic of length l , then

(12.4.2) $\quad E(u(\gamma,g)) \geq c/l$,

as $l \to 0$, where $u(\gamma,g)$ is a harmonic map homotopic to the identity. By the collar theorem of Keen [Ke] and Halpern [Hp], (Σ,γ) contains an isometric copy of the region $\{z: 1 < |z| < e^l , \theta < \arg z < \pi - \theta\}$ in the Poincaré upper half plane, identifying $|z| = 1$ and $|z| = e^l$ via $z \to e^l z$, and $\theta \to 0$ as $l \to 0$.
Since the homotopically nontrivial closed curves on (Σ,g) have length $\geqq l_o > 0$, we have

$$(12.4.3) \qquad \int_1^{e^l} (g_{ij} \frac{\partial u^i}{\partial s} \frac{\partial u^j}{\partial s})^{1/2} ds \geqq l_o ,$$

where s denotes arclength along a curve $\arg z = \varphi$. Noting $(g_{ij} \frac{\partial u^i}{\partial s} \frac{\partial u^j}{\partial s}) \leq 2e(u)$, $ds = (r \sin\varphi)^{-1} dr$, and using Hölder's inequality, (12.4.3) implies

$$l_o^2 \leq (\int_1^{e^l} \frac{dr}{r})(\int_1^{e^l} 2e(u) \frac{dr}{r \sin^2\varphi}) = l \int_1^{e^l} 2e(u) \frac{dr}{r \sin^2\varphi}$$

and integrating w.r.t. φ

$$\frac{l_o^2(\pi - 2\theta)}{l} \leq \int_\theta^{\pi-\theta} \int_1^{e^l} 2e(u) \frac{dr d\varphi}{r \sin^2\varphi} \leq 2E(u) .$$

Since, as noted above, $\theta \to 0$ as $l \to 0$, (12.4.2) follows.
By Thm. 11.1, for any metric γ on Σ , there exists a harmonic map $u(\gamma,g)$, homotopic to the identity of Σ . By Lemma 6.1, we can assume w.l.o.g. that γ has constant curvature -1 . If we now minimize w.r.t. γ , (12.4.2) allows us to assume that a minimizing sequence $u_i(\gamma_i,g)$ stays within a region

$\{p \in T(\Sigma)$: the length of the shortest closed geodesic with respe to the constant curvature metric representing p is bounded below by $c_o > 0\}$.

By a Theorem of Mumford [Mu], such a set is a compact subset of the moduli space $R(\Sigma) = M(\Sigma)/D(\Sigma)$. Suppose on the other hand that $u_i(\gamma_i,g)$ have bounded energy, where γ_i is a sequence of mutually distinct points in $T(\Sigma)$ which all correspond to the same point $\tilde{\gamma}$ in $R(\Sigma)$ under the natural projection $T(\Sigma) \to R(\Sigma) = T(\Sigma)/$

$(D(\Sigma)/D_o(\Sigma))$. By equicontinuity of the $u_i(\gamma_i,g)$ (cf. 3.3 or 11.2), using the fact that all γ_i have the same injectivity radius for their constant curvature metric), the $u_i(\gamma_i,g)$ have a uniformly convergent subsequence which contradicts the fact that all u_i (γ_i,g) are in different homotopy classes[1)] by choice of γ_i. These assertions imply that E is a proper function on $T(\Sigma)$ and that a minimizing sequence $u_i(\gamma_i,g)$ stays within a compact region of $T(\Sigma)$. By equicontinuity again (by Mumford's Theorem, all γ_i now have injectivity radius uniformly bounded below for their constant curvature metric), we can choose a convergent subsequence $\gamma_i \to \gamma$, $u_i(\gamma_i,g) \to u(\gamma,g)$. u is harmonic with respect to γ an minimizing by continuity of E as a function of γ.

1) when composed with the conformal and hence energy preserving map $\gamma_1 \to \gamma_i$

<u>Theorem 12.6</u>: $T(\Sigma)$ <u>is topologically a cell.</u>

<u>Proof</u>: (A. Tromba [Tr]):
We fix any conformal structure p on Σ and represent it by some metric g with nonpositive curvature. For every metric γ, we can find a harmonic map $u(\gamma,g)$ homotopic to the identity. We now look at $E = E(u(\gamma,g))$ as a function of the conformal structure represented by γ. Every critical point gives rise to a conformal branched cover $u(\gamma,g)$, by Thm. 12.4, and since $u(\gamma,g)$ is homotopic to the identity and therefore of degree one, no branch points occur.
Therefore, the global minimum p of Thm. 12.5 has to be the only critical point of E on $T(\Sigma)$.
It is straightforward to calculate that the second variation of E at p is positive definite (cf. [Tr]). Hence p is a nondegenerate minimum, and the result follows from Morse theory.

12.5. The approach of Gerstenhaber - Rauch

Still another application of harmonic maps to Teichmüller theory was outlined by Gerstenhaber and Rauch [GR] in 1954. They tried to obtain Teichmüller's extremal quasiconformal maps by solving integral variational problems. Given two complex structures on a surface, they minimize energy in a given homotopy class of maps between these structures for any conformal metric on the image and then maximize the energy of the corresponding harmonic map over all such conformal metrics with fixed area and finally assume a continuous dependance of the minimizing map on the metric to show that a result of such a twofold variational procedure is an extremal quasiconformal map. While the solution of the first variational problem is provided by Lemaire's Theorem, the existence of a solution to the second one is not clear, and the continuous dependance is also not known because uniqueness of harmonic maps is not known for an arbitrarily curved image metric, cf. 5.5. Therefore, this is an interesting, but unfortunately still incomplete program.

12.6. Harmonic Gauss maps and Bernstein theorems

Let $F: X \to \mathbb{R}^{n+p}$ denote an immersion of the n - dimensional manifold X into $\mathbb{R}^{n+p}$. The Gauss map $G: X \to G(p,n)$ into the Grassmannian manifold $G(p,n)$ of p - planes in $\mathbb{R}^{n+p}$ assigns to each point $x \in X$ the normal plane at $F(x)$.
Ruh and Vilms [RV] proved

Theorem 12.7: $G: X \to G(p,n)$ is harmonic, if and only if X is immersed into $\mathbb{R}^{n+p}$ with parallel mean curvature field.

Corollary 12.4: Suppose Σ is a surface of constant mean curvature H in $\mathbb{R}^3$. Then the Gauss map $G: X \to S^2$ is harmonic.

If $H = 0$ in Cor. 12.4, i.e. Σ is a minimal surface, then the Gauss map actually is antiholomorphic. These results can be used to prove Bernstein type theorems for minimal submanifolds or submanifolds with parallel mean curvature field of Euclidean space. This approach proves a Liouville type theorem for harmonic maps into Grassmannian manifolds under certain size restrictions on the image. This means that the Gauss map is shown to be constant which in turn implies that the corresponding immersed submanifold is a linear subspace of $\mathbb{R}^{n+p}$.
Here, we do not want to elaborate this point, but refer the reader instead to [HJW] and [HOS] for some results in this direction.

12.7. Surfaces of constant Gauss curvature in 3 - space

Another example arises from (immersed) surfaces of constant Gauss curvature in 3 - space.
Suppose F is a surface with local coordinates u,v. If the Gauss curvature K of F is positive, then the second fundamental form

$$II: = h_{11}du^2 + 2h_{12}dudv + h_{22}dv^2$$

is positive definite and can hence be diagonalized by introducing new parameters x,y instead of u,v (i.e. $II = \lambda(dx^2 + dy^2)$, where λ is a positive function). x,y are called isothermal conjugated parameters, and Darboux ([Db], §725) discovered that u and v as functions of x and y satisfy a system of elliptic equations depending only on the first fundamental form of F, i.e. only on intrinsic geometric quantities, not depending on the immersion of F into 3-space. This fact was used extensively by H. Lewy [Lw] and E. Heinz ([Hz2]) for solving the Weyl embedding problem.

The system discovered by Darboux is the following

$$(12.7.1) \qquad \Delta u + (\Gamma^1_{11} + 1/2 \frac{\partial}{\partial u} (\log K))(u_x^2 + u_y^2) + (2\Gamma^1_{12} + 1/2 \frac{\partial}{\partial v} (\log K))(u_x v_x + u_y v_y) + \Gamma^1_{22} (v_x^2 + v_y^2) = 0$$

$$\Delta v + \Gamma^2_{11} (u_x^2 + u_y^2) + (2\Gamma^2_{12} + 1/2 \frac{\partial}{\partial u} (\log K))(u_x v_x + u_y v_y) + (\Gamma^2_{22} + 1/2 \frac{\partial}{\partial v} (\log K))(v_x^2 + v_y^2) = 0$$

We see in particular that, if the Gauss curvature K of F is a positive constant, then the transformation $(x,y) \to (u,v)$ is harmonic. (This fact was pointed out to me by S. Hildebrandt).

As an application, we present a proof of the following well-known theorem of Liebmann (cf. [Bl], p. 195).

Theorem 12.8: The only immersion of a closed surface F of constant positive curvature K into 3-space is given by a standard sphere of radius $1/\sqrt{K}$.

Proof: By the Gauss-Bonnet Theorem, F is topologically a sphere. Since the second fundamental form of a given immersion of F is positive definite, we can use the uniformization theorem to obtain parameters x,y on the sphere S^2 which diagonalize this form. Since $K \equiv$ const., we thus obtain a harmonic map

$$h: S^2 \to F .$$

By Cor. 12.1, h is (anti) conformal and therefore also diagonalizes the first fundamental form. Hence the first and the second fundamental form are proportional everywhere, which means that the given immersion is everywhere umbilical and therefore a standard sphere (cf. [Bl] p.97).

If we note that in the sphere - case, Cor. 12.1 follows from Lemma 1.1 and the Theorem of Riemann - Roch, then we see that the preceding proof is much in the spirit of H. Hopf's proof, that every immersion of a sphere into 3 - space with constant mean curvature is a standard sphere (cf. [Ho]).

References .

[Ad] Adams, R., Sobolev Spaces, Academic Press, 1975

[A1] Ahlfors, L., Some Remarks on Teichmüller's Space of Riemann Surfaces, Ann. Math. 74 (1961), 171 - 191

[A2] Ahlfors, L., Curvature Properties of Teichmüller's Space, J. d'Anal. Math. 9 (1961), 161 - 176

[AB] Ahlfors, L., and L. Bers, Riemann's Mapping Theorem for Variable Metrics, Ann. Math. 72 (1960), 385 - 404

[Bg] Berg, P., On Univalent Mappings by Solutions of Linear Partial Differential Equations, Trans. Amer. Math. Soc. (1957), 310 - 318

[B1] Bers, L., Quasiconformal Mappings and Teichmüller's Theorem, in: Analytic Functions, Princeton Univ. Press, Princeton, (1960), 89 - 119

[B2] Bers, L., An Outline of the Theory of Pseudoanalytic Functions, Bull. Amer. Math. Soc. 62 (1956), 291 - 331

[BJS] Bers, L., F. John, and M. Schechter, Partial Differential Equations, Interscience, New York, (1964)

[Bl] Blaschke, W., Vorlesungen über Differentialgeometrie, part I, Springer, Berlin, 4 1945

[BC1] Brezis, H., and J. M. Coron, Multiple Solutions of H - systems and Rellich's Conjecture, Comm. Pure Appl. Math., to appear

[BC2] Brezis, H., and J. M. Coron, Large solutions for harmonic maps in two dimensions, Comm. Math. Phys., to appear

[Ca] Calabi, E., An Intrinsic Characterization of Harmonic One - Forms, in: Global Analysis, ed. by D. C. Spencer and S. Iyanaga, Princeton Univ. Press, Tokyo, Princeton, (1969)

[Ci] Choi, H. J., On the Liouville Theorem for Harmonic Maps, Preprint

[Cq] Choquet, G., Sur un type de transformation analytique généralisant la répresentation conforme et définie au moyen de fonctions harmoniques, Bull. Sci. Math. (2) 69 (1945), 156 - 165

[Co] Courant, R., Dirichlet's Principle, Conformal Mapping, and Minimal Surfaces, New York, Interscience 1950

[Db] Darboux, G., Théorie générale des surfaces, Tome III, Paris, 1894

[Dl] Deimling, K., Nichtlineare Gleichungen und Abbildungsgrade, Springer, Berlin, Heidelberg, New York, 1974

[DTK] De Turck, D., and J. Kazdan, Some Regularity Theorems in Riemannian Geometry, Ann. Sc. Ec. N. Sup. Paris

[EE] Earle, C. J., and J. Eells, A Fibre Bundle Description of Teichmüller Theory, J. Diff. Geom. 3 (1969), 19 - 43

[ESz] Earle, C. J., and A. Schatz, Teichmüller Theory for Surfaces with Boundary, J. Diff. Geom. 4 (1970), 169 - 185

[E] Eells, J., Regularity of Certain Harmonic Maps, Proc. Durham Conf. 1982

[EL1] Eells, J., and L. Lemaire, A Report on Harmonic Maps, Bull. London Math. Soc. 10, 1 - 68 (1978)

[EL2] Eells, J., and L. Lemaire, Deformations of Metrics and Associated Harmonic Maps, Patodi, Mem. Vol. Geometry and Analysis Tata Inst. 1980, 33 - 45

[EL3] Eells, J., and L. Lemaire, On the Construction of Harmonic an Holomorphic Maps Between Surfaces, Math. Ann 252 (1980), 27 - 52

[EL4] Eells, J., and L. Lemaire, Selected Topics in Harmonic Maps, CBMS Regional Conf. 1981

[ES] Eells, J., and J. H. Sampson, Harmonic Mappings of Riemannian Manifolds, Am. J. Math. 86 (1964), 109 - 160

[EW1] Eells, J., and J. C. Wood, Restrictions on Harmonic Maps of Surfaces, Top. 15 (1976), 263-266

[EW2] Eells, J., and J. C. Wood, Harmonic Maps from Surfaces to Complex Projective Spaces, Preprint, Univ. Warwick, 1981

[EW3] Eells, J., and J. C. Wood, The Existence and Construction of Certain Harmonic Maps, Ist. Francesco Severi, Symp. Math. 26, (1982), 123 - 138

[Es] Eliasson, H. I., A Priori Growth and Hölder Estimates for Harmonic Mappings, Univ. Iceland, Preprint, 1981

[F] Federer, H., Geometric Measure Theory, Springer, Grundlehren 153, New York, 1969

[Fe] Fenchel, W., Elementare Beweise und Anwendungen einiger Fixpunktsätze, Mat. Tidsskr. (B) (1932), 66 - 87

[FT] Fischer, E., and A. Tromba, On a Purely "Riemannian" Proof of the Structure and Dimension of the Unramified Moduli Space of a Compact Riemann Surface, SFB 72, Preprint 516, Bonn, 1982

[GR] Gerstenhaber, M., and H. E. Rauch, On Extremal Quasiconformal Mappings I, II, Proc. Nat. Ac. Sc. 40 (1954), 808 - 812 and 991 - 994

[G] Giaquinta, M., Multiple Integrals in the Calculus of Variations and Non Linear Elliptic Systems, SFB 72, Vorlesungsreih No. 6, Bonn, 1981

[GG1] Giaquinta, M., and Giusti, E., On the Regularity of the Mini of Variational Integrals, Acta Math. 148 (1982) 31 - 46

[GG2] Giaquinta, M., and Giusti, E., The Singular Set of the Minim of Certain Quadratic Functionals, to appear in Analysis

[GH] Giaquinta, M., and S. Hildebrandt, A Priori Estimates for Ha monic Mappings, Journ. Reine Angew. Math.

[GT] Gilbarg, D., and N. S. Trudinger, Elliptic Partial Differential Equations of Second Order, Springer, Grundlehren 224, Berlin, Heidelberg, New York, 1977

[Go] Gordon, W., Convex Functions and Harmonic Mappings, Proc. A. M. S. 33 (1972), 433 - 437

[GKM] Gromoll, D., W. Klingenberg, and W. Meyer, Riemannsche Geometrie im Großen, L.N.M. 55, Springer, Berlin, Heidelberg, New York, 21975

[GJ] Gulliver, R., and J. Jost, Harmonic Maps which Solve a Free Boundary Value Problem, in preparation

[Hp] Halpern, N., A Proof of the Collar Lemma, Bull. London. Math. Soc. 13 (1981), 141 - 144

[Hm] Hamilton, R., Harmonic Maps of Manifolds with Boundary, L.N. M. 471, Springer, Berlin, Heidelberg, New York, 1975

[Ht] Hartman, P., On Homotopic Harmonic Maps, Can. J. Math. 19 (1967) 673 - 687

[HtW] Hartman, P., and A. Wintner, On the Local Behavior of Solutions of Nonparabolic Partial Differential Equations, Amer. J. Math. 75 (1953), 449 - 476

[Hz1] Heinz, E., On Certain Nonlinear Elliptic Differential Equations and Univalent Mappings, Journ. d'Anal. 5 (1956/57) 197 - 272

[Hz2] Heinz, E., Neue a - priori Abschätzungen für den Ortsvektor einer Fläche positiver Gaußscher Krümmung durch ihr Linienelement, Math. Z. 74, (1960), 129 - 157

[Hz3] Heinz, E., Existence Theorems for One - to - One Mappings Associated with Elliptic Systems of Second Order, part I, Journ. d'Anal. 15 (1965) 325 - 353

[Hz4] Heinz, E., Existence Theorems for One - to - One Mappings Associated with Elliptic Systems of Second Order, part II, Journ. d'Anal 17, (1967), 145 - 184

[Hz5] Heinz, E., Über das Nichtverschwinden der Funktionaldeterminante bei einer Klasse eineindeutiger Abbildungen, M. Z. 105, (1968), 87 - 89

[Hz6] Heinz, E., Zur Abschätzung der Funktionaldeterminante bei einer Klasse topologischer Abbildungen, Nachr. Akad. Wiss. Gött. (1968), 183 - 197

[Hz7] Heinz, E., Über das Randverhalten quasilinearer elliptischer Systeme mit isothermen Parametern, M. Z. 113, (1970), 99 - 105

[Hi1] Hildebrandt, S., On the Plateau Problem for Surfaces of Constant Mean Curvature, Comm. Pure Appl. Math. 23, (1970), 97 - 114

[Hi2] Hildebrandt, S., Nonlinear Elliptic Systems and Harmonic Mappings, Proj. Beijing Symp. Diff. Geom. & Diff. Eq. 1980, Science Press, Beijing, 1982, also in SFB 72, Vorlesungsreihe No. 3, Bonn, 1980

[HJW] Hildebrandt, S., J. Jost, and K.-O. Widman, Harmonic Mappings and Minimal Submanifolds, Inv. math. 62 (1980), 269-298

[HKW1] Hildebrandt, S., H. Kaul, and K.-O. Widman, Harmonic Mappings into Riemannian Manifolds with Non-positive Sectional Curvature, Math. Scand. 37 (1975), 257-263

[HKW2] Hildebrandt, S., H. Kaul, and K.-O. Widman, Dirichlet's Boundary Value Problem for Harmonic Mappings of Riemannian Manifolds, M. Z. 147 (1976), 225-236

[HKW3] Hildebrandt, S., H. Kaul and K.-O. Widman, An Existence Theorem for Harmonic Mappings of Riemannian Manifolds, Acta Math. 138 (1977) 1-16

[HW1] Hildebrandt, S., and K.-O. Widman, Some Regularity Results for Quasilinear Elliptic Systems of Second Order, Math. Z. 142 (1975), 67-86

[HW2] Hildebrandt, S., and K.-O. Widman, On the Hölder Continuity of Weak Solutions of Quasilinear Elliptic Systems of Second Order, Ann. Sc. N. Sup. Pisa IV (1977), 145-178

[HOS] Hoffman, D. A., R. Osserman, and R. Schoen, On the Gauss Map of Complete Surfaces of Constant Mean Curvature in $\mathbb{R}^3$ and $\mathbb{R}^4$, Comm. Math. Helv. 57 (1982), 519 - 531

[Ho] Hopf, H., Über Flächen mit einer Relation zwischen den Hauptkrümmungen, Math. Nachr. 4 (1950/51), 232-249

[Ih] Ishihara, T., A Mapping of Riemannian Manifolds which Preserves Harmonic Functions, J. Math. Kyoto Univ. 19 (1979), 215-229

[JäK1] Jäger, W., and H. Kaul, Uniqueness of Harmonic Mappings and of Solutions of Elliptic Equations on Riemannian Manifolds, Math. Ann. 240 (1979), 231-250

[JäK2] Jäger, W., and H. Kaul, Uniqueness and Stability of Harmonic Maps and their Jacobi Fields, Man. Math. 28, (1979), 269-291

[J1] Jost, J., Eineindeutigkeit harmonischer Abbildungen, Diplomarbeit, Bonn, 1979, also Bonner Math. Schr. 129 (1981)

[J2] Jost, J., Eine geometrische Bemerkung zu Sätzen über harmonische Abbildungen, die ein Dirichletproblem lösen, Man. math. 32, (1980) 51-57

[J3] Jost, J., Univalency of Harmonic Mappings between Surfaces, Journ. Reine Angew. Math. 324, (1981) 141 - 153

[J4] Jost, J., Ein Existenzbeweis für harmonische Abbildungen, die ein Dirichletproblem lösen, mittels der Methode des Wärmeflusses, Man. math. 34 (1981), 17-25

[J5] Jost, J., A Maximum Principle for Harmonic Mappings which solve a Dirichlet Principle, man math. 38 (1982), 129-130

[J6] Jost, J., Existence Proofs for Harmonic Mappings with the Help of a Maximum Principle, M. Z. 184 (1983), 489 - 496

J7] Jost, J., The Dirichlet Problem for Harmonic Maps from a Surface with Boundary onto a 2 - Sphere with Non-Constant Boundary Values, Preprint

JK1] Jost, J., and H. Karcher, Geometrische Methoden zur Gewinnung von a - priori - Schranken für harmonische Abbildungen, man. math. 40 (1982), 27 - 77

JK2] Jost, J., and H. Karcher, Almost Linear Functions and a-priori Estimates for Harmonic Maps, Proc. Durham. Conf. 1982

JM] Jost, J., and M. Meier, Boundary Regularity for Minima of Certain Quadratic Functionals, Math. Ann. 262 (1983), 549 - 561

JS] Jost, J., and R. Schoen, On the Existence of Harmonic Diffeomorphisms Between Surfaces, Inv. math. 66 (1982), 353 - 359

JY] Jost, J., and S. T. Yau, Harmonic Mappings and Kähler Manifolds, Math. Ann. 262 (1983), 145 - 166

K1] Karcher, H., Schnittort und konvexe Mengen in vollständigen Riemannschen Mannigfaltigkeiten, Math. Ann. 177 (1968), 105 - 121

K2] Karcher, H., Riemannian Center of Mass and Mollifier Smoothing, CPAM 30, (1977) 509 - 541

KW] Karcher,H., and J.C. Wood, Non-Existence Results and Growth Properties for Harmonic Maps and Forms, SFB-40 Preprint, Bonn, 1983

l1] Kaul, H., Schranken für die Christoffelsymbole, man. math. 19 (1976) 261 - 273

e] Keen, L., Collars on Riemann Surfaces, Ann. Math. Studies 79 (1974), 263 - 268

n1] Kneser, H., Lösung der Aufgabe 41, Jber. Dtsch. Math. Ver. 35 (1926), 123 - 124

n2] Kneser, H., Die kleinste Bedeckungszahl innerhalb einer Klasse von Flächenabbildungen, Math. Ann. 103 (1930), 347 - 358

U] Ladyženskaja, O. A., and N. N. Ural'ceva, Équations aux dérivées partielles de type elliptique, Dunod, Paris 1968

v] Lavrent'ev, M. A., Sur une classe des représentations continues, Mat. Sb. 42 (1935), 407 - 434

z] Leibniz, G. W., Die Theodizee, Phil. Bibl. 71, Felix Meiner, Hamburg, ²1968

1] Lemaire, L., Applications harmoniques de surfaces Riemanniennes, J. Diff. Geom. 13 (1978), 51 - 78

2] Lemaire, L., Boundary Value Problems for Harmonic and Minimal Maps of Surfaces into Manifolds, Ann. Sc. Norm. Sup.Pisa (4) 8 (1982), 91 - 103

S] Leray, J., and J. Schauder, Topologie et équations fonctionnelles, Ann. École Norm. Sup. 51 (1934), 45 - 78

[Lw] Lewy, H., On the Existence of a Closed Surface Realizing a Given Riemannian Metric, Proc., N.A.S. USA 24, 2 (1938), 104 - 106

[Li] Lichtenstein, L., Zur Theorie der konformen Abbildung. Konforme Abbildung nichtanalytischer singularitätenfreier Flächenstücke auf ebene Gebiete, Bull. Acad. Sci. Cracovie, Cl. Sci. Mat. Nat. A (1916), 192 - 217

[MS] Meyers, N., and J. Serrin, H = W, Proc. Nat. Ac. Sc. 51 (1964 1055 - 1056

[Mi] Misner, Harmonic Maps as Models for Physical Theories, Phys. Rev. D 18 (12) (1978)

[M1] Morrey, C. B., On the Solutions of Quasi - Linear Elliptic Partial Differential Equations, Trans. A.M.S. 43 (1938), 126 - 166

[M2] Morrey, C. B., The Problem of Plateau on a Riemannian Manifold, Ann. of Math. 49 (1948), 807 - 851

[M3] Morrey, C. B., Multiple Integrals in the Calculus of Variations, Springer, Berlin, Heidelberg, New York, 1966

[Mu] Mumford, D., A Remark on Mahler's Compactness Theorem, Proc. A.M.S. 28 (1971), 289 - 294

[Na] Nash, J., The Embedding Problem for Riemannian Manifolds, Ann Math. 63 (1956), 20 - 63

[O] Olivier, R., Die Existenz geschlossener Geodätischer auf kompakten Mannigfaltigkeiten, Comm. Math. Helv. 35 (1961), 146 - 152

[Rd] Radó, T., Aufgabe 41, Jber. Dtsch. Math. Ver. 35 (1926), 49

[RV] Ruh, E. A., and J. Vilms, The Tension Field of the Gauss Map, Trans. A.M.S. 149 (1970), 569 - 573

[SkU] Sacks, J. and K. Uhlenbeck, The Existence of Minimal Immersions of 2 - Spheres, Ann. Math. 113 (1981), 1 - 24

[Sa] Sampson, J. H., Some Properties and Applications of Harmonic Mappings, Ann. Sc. Ec. Sup. 11 (1978), 211 - 228

[SU1] Schoen, R., and K. Uhlenbeck, A regularity theory for harmonic maps, J. Diff. Geom. 17 (1982), 307 - 335

[SU2] Schoen, R., and K. Uhlenbeck, Boundary regularity and miscellaneous results on harmonic maps, to appear in J. Diff. Geom

[SY1] Schoen, R., and S. T. Yau, On Univalent Harmonic Maps betwee Surfaces, Inv. math. 44 (1978), 265 - 278

[SY2] Schoen, R., and S. T. Yau, Existence of Incompressible Minimal Surfaces and the Topology of Three Dimensional Manifolds with Non-Negative Scalar Curvature, Ann. Math. 110 (1979), 127 - 142

[SY3] Schoen, R., and S. T. Yau, Compact Group Actions and the Topology of Manifolds with Non-Positive Curvature, Top. 18, (1979), 361 - 380

[Se1] Sealey, H., Some Properties of Harmonic Mappings, Thesis, Univ. Warwick, 1980

[Se2] Sealey, H., The Stress - Energy Tensor and the Vanishing of L^2-Harmonic Forms, Preprint

[Sh] Shibata, K., On the Existence of a Harmonic Mapping, Osaka J. Math. 15 (1963), 173 - 211

[Si] Siu, Y. T., The Complex Analyticity of Harmonic Maps and the Strong Rigidity of Compact Kähler Manifolds, Ann. Math. 112 (1980), 73 - 111

[Sp] Sperner, E., A priori Gradient Estimates for Harmonic Mappings, SFB 72, Preprint 513, Bonn, 1982

[St] Stampacchia, G., Le problème de Dirichlet pour les équations elliptiques du second ordre à coefficients discontinues, Ann. Inst. Fourier (Grenoble) 15, 189 - 258 (1965)

[Tm] Teichmüller, O., Extremale quasikonforme Abbildungen und quadratische Differentiale, Abh. Preuss. Akad. Wiss. Math. - Nat. Kl. 22 (1939)

[Td] Tolksdorf, P., A Strong Maximum Principle and Regularity for Harmonic Mappings, Preprint

[Tr] Tromba, A., A New Proof that Teichmüller Space is a Cell, Preprint

[U] Uhlenbeck, K., Harmonic Maps: A Direct Method in the Calculus of Variations, Bull. AMS 76 (1970), 1082 - 1087

[Wt] Wente, H., The Differential Equations $\Delta x = 2Hx_u \wedge x_v$ with Vanishing Boundary Values, Proc. A.M.S. 50 (1975), 131 - 137

[Wy] Weyl, H., Die Idee der Riemannschen Fläche, Teubner, Leipzig, 1955

[Wi] Wiegner, M., A priori Schranken für Lösungen gewisser elliptischer Systeme, Man. math. 18 (1976), 279 - 297

[W1] Wood, J. C., Singularities of Harmonic Maps and Applications of the Gauss - Bonnet Formula, Amer. J. Math. 99 (1977), 1329 - 1344

[W2] Wood, J. C., Non - existence of Solutions to Certain Dirichlet Problems, Preprint, Leeds, 1981.

. 900: P. Deligne, J. S. Milne, A. Ogus, and K.-Y. Shih, Hodge cles, Motives, and Shimura Varieties. V, 414 pages. 1982.

. 901: Séminaire Bourbaki vol. 1980/81 Exposés 561-578. III, 9 pages. 1981.

. 902: F. Dumortier, P.R. Rodrigues, and R. Roussarie, Germs Diffeomorphisms in the Plane. IV, 197 pages. 1981.

. 903: Representations of Algebras. Proceedings, 1980. Edited M. Auslander and E. Lluis. XV, 371 pages. 1981.

. 904: K. Donner, Extension of Positive Operators and Korovkin orems. XII, 182 pages. 1982.

. 905: Differential Geometric Methods in Mathematical Physics. ceedings, 1980. Edited by H.-D. Doebner, S.J. Andersson, and R. Petry. VI, 309 pages. 1982.

906: Séminaire de Théorie du Potentiel, Paris, No. 6. Proceeds. Edité par F. Hirsch et G. Mokobodzki. IV, 328 pages. 1982.

. 907: P. Schenzel, Dualisierende Komplexe in der lokalen ebra und Buchsbaum-Ringe. VII, 161 Seiten. 1982.

908: Harmonic Analysis. Proceedings, 1981. Edited by F. Ricci G. Weiss. V, 325 pages. 1982.

909: Numerical Analysis. Proceedings, 1981. Edited by J.P. nart. VII, 247 pages. 1982.

910: S.S. Abhyankar, Weighted Expansions for Canonical Degularization. VII, 236 pages. 1982.

. 911: O.G. Jørsboe, L. Mejlbro, The Carleson-Hunt Theorem on rier Series. IV, 123 pages. 1982.

912: Numerical Analysis. Proceedings, 1981. Edited by G. A. tson. XIII, 245 pages. 1982.

913: O. Tammi, Extremum Problems for Bounded Univalent ctions II. VI, 168 pages. 1982.

914: M. L. Warshauer, The Witt Group of Degree k Maps and nmetric Inner Product Spaces. IV, 269 pages. 1982.

915: Categorical Aspects of Topology and Analysis. Proceed-, 1981. Edited by B. Banaschewski. XI, 385 pages. 1982.

916: K.-U. Grusa, Zweidimensionale, interpolierende Lg-Splines ihre Anwendungen. VIII, 238 Seiten. 1982.

917: Brauer Groups in Ring Theory and Algebraic Geometry. Prodings, 1981. Edited by F. van Oystaeyen and A. Verschoren. VIII, pages. 1982.

918: Z. Semadeni, Schauder Bases in Banach Spaces of tinuous Functions. V, 136 pages. 1982.

919: Séminaire Pierre Lelong – Henri Skoda (Analyse) Années /81 et Colloque de Wimereux, Mai 1981. Proceedings. Edité P. Lelong et H. Skoda. VII, 383 pages. 1982.

920: Séminaire de Probabilités XVI, 1980/81. Proceedings. par J. Azéma et M. Yor. V, 622 pages. 1982.

921: Séminaire de Probabilités XVI, 1980/81. Supplément: Géoe Différentielle Stochastique. Proceedings. Edité par J. Azéma Yor. III, 285 pages. 1982.

922: B. Dacorogna, Weak Continuity and Weak Lower Seminuity of Non-Linear Functionals. V, 120 pages. 1982.

923: Functional Analysis in Markov Processes. Proceedings, Edited by M. Fukushima. V, 307 pages. 1982.

924: Séminaire d'Algèbre Paul Dubreil et Marie-Paule Malliavin. eedings, 1981. Edité par M.-P. Malliavin. V, 461 pages. 1982.

925: The Riemann Problem, Complete Integrability and Arith- Applications. Proceedings, 1979-1980. Edited by D. Chudnovnd G. Chudnovsky. VI, 373 pages. 1982.

926: Geometric Techniques in Gauge Theories. Proceedings, Edited by R. Martini and E.M.de Jager. IX, 219 pages. 1982.

Vol. 927: Y. Z. Flicker, The Trace Formula and Base Change for GL (3). XII, 204 pages. 1982.

Vol. 928: Probability Measures on Groups. Proceedings 1981. Edited by H. Heyer. X, 477 pages. 1982.

Vol. 929: Ecole d'Été de Probabilités de Saint-Flour X – 1980. Proceedings, 1980. Edited by P.L. Hennequin. X, 313 pages. 1982.

Vol. 930: P. Berthelot, L. Breen, et W. Messing, Théorie de Dieudonné Cristalline II. XI, 261 pages. 1982.

Vol. 931: D.M. Arnold, Finite Rank Torsion Free Abelian Groups and Rings. VII, 191 pages. 1982.

Vol. 932: Analytic Theory of Continued Fractions. Proceedings, 1981. Edited by W.B. Jones, W.J. Thron, and H. Waadeland. VI, 240 pages. 1982.

Vol. 933: Lie Algebras and Related Topics. Proceedings, 1981. Edited by D. Winter. VI, 236 pages. 1982.

Vol. 934: M. Sakai, Quadrature Domains. IV, 133 pages. 1982.

Vol. 935: R. Sot, Simple Morphisms in Algebraic Geometry. IV, 146 pages. 1982.

Vol. 936: S.M. Khaleelulla, Counterexamples in Topological Vector Spaces. XXI, 179 pages. 1982.

Vol. 937: E. Combet, Intégrales Exponentielles. VIII, 114 pages. 1982.

Vol. 938: Number Theory. Proceedings, 1981. Edited by K. Alladi. IX, 177 pages. 1982.

Vol. 939: Martingale Theory in Harmonic Analysis and Banach Spaces. Proceedings, 1981. Edited by J.-A. Chao and W.A. Woyczyński. VIII, 225 pages. 1982.

Vol. 940: S. Shelah, Proper Forcing. XXIX, 496 pages. 1982.

Vol. 941: A. Legrand, Homotopie des Espaces de Sections. VII, 132 pages. 1982.

Vol. 942: Theory and Applications of Singular Perturbations. Proceedings, 1981. Edited by W. Eckhaus and E.M. de Jager. V, 363 pages. 1982.

Vol. 943: V. Ancona, G. Tomassini, Modifications Analytiques. IV, 120 pages. 1982.

Vol. 944: Representations of Algebras. Workshop Proceedings, 1980. Edited by M. Auslander and E. Lluis. V, 258 pages. 1982.

Vol. 945: Measure Theory. Oberwolfach 1981, Proceedings. Edited by D. Kölzow and D. Maharam-Stone. XV, 431 pages. 1982.

Vol. 946: N. Spaltenstein, Classes Unipotentes et Sous-groupes de Borel. IX, 259 pages. 1982.

Vol. 947: Algebraic Threefolds. Proceedings, 1981. Edited by A. Conte. VII, 315 pages. 1982.

Vol. 948: Functional Analysis. Proceedings, 1981. Edited by D. Butković, H. Kraljević, and S. Kurepa. X, 239 pages. 1982.

Vol. 949: Harmonic Maps. Proceedings, 1980. Edited by R.J. Knill, M. Kalka and H.C.J. Sealey. V, 158 pages. 1982.

Vol. 950: Complex Analysis. Proceedings, 1980. Edited by J. Eells. IV, 428 pages. 1982.

Vol. 951: Advances in Non-Commutative Ring Theory. Proceedings, 1981. Edited by P.J. Fleury. V, 142 pages. 1982.

Vol. 952: Combinatorial Mathematics IX. Proceedings, 1981. Edited by E. Billington, S. Oates-Williams, and A.P. Street. XI, 443 pages. 1982.

Vol. 953: Iterative Solution of Nonlinear Systems of Equations. Proceedings, 1982. Edited by R. Ansorge, Th. Meis, and W. Törnig. VII, 202 pages. 1982.

Vol. 954: S.G. Pandit, S.G. Deo, Differential Systems Involving Impulses. VII, 102 pages. 1982.

Vol. 955: G. Gierz, Bundles of Topological Vector Spaces and Their Duality. IV, 296 pages. 1982.

Vol. 956: Group Actions and Vector Fields. Proceedings, 1981. Edited by J.B. Carrell. V, 144 pages. 1982.

Vol. 957: Differential Equations. Proceedings, 1981. Edited by D.G. de Figueiredo. VIII, 301 pages. 1982.

Vol. 958: F.R. Beyl, J. Tappe, Group Extensions, Representations, and the Schur Multiplicator. IV, 278 pages. 1982.

Vol. 959: Géométrie Algébrique Réelle et Formes Quadratiques, Proceedings, 1981. Edité par J.-L. Colliot-Thélène, M. Coste, L. Mahé, et M.-F. Roy. X, 458 pages. 1982.

Vol. 960: Multigrid Methods. Proceedings, 1981. Edited by W. Hackbusch and U. Trottenberg. VII, 652 pages. 1982.

Vol. 961: Algebraic Geometry. Proceedings, 1981. Edited by J.M. Aroca, R. Buchweitz, M. Giusti, and M. Merle. X, 500 pages. 1982.

Vol. 962: Category Theory. Proceedings, 1981. Edited by K.H. Kamps, D. Pumplün, and W. Tholen, XV, 322 pages. 1982.

Vol. 963: R. Nottrot, Optimal Processes on Manifolds. VI, 124 pages. 1982.

Vol. 964: Ordinary and Partial Differential Equations. Proceedings, 1982. Edited by W.N. Everitt and B.D. Sleeman. XVIII, 726 pages. 1982.

Vol. 965: Topics in Numerical Analysis. Proceedings, 1981. Edited by P.R. Turner. IX, 202 pages. 1982.

Vol. 966: Algebraic K-Theory. Proceedings, 1980, Part I. Edited by R.K. Dennis. VIII, 407 pages. 1982.

Vol. 967: Algebraic K-Theory. Proceedings, 1980. Part II. VIII, 409 pages. 1982.

Vol. 968: Numerical Integration of Differential Equations and Large Linear Systems. Proceedings, 1980. Edited by J. Hinze. VI, 412 pages. 1982.

Vol. 969: Combinatorial Theory. Proceedings, 1982. Edited by D. Jungnickel and K. Vedder. V, 326 pages. 1982.

Vol. 970: Twistor Geometry and Non-Linear Systems. Proceedings, 1980. Edited by H.-D. Doebner and T.D. Palev. V, 216 pages. 1982.

Vol. 971: Kleinian Groups and Related Topics. Proceedings, 1981. Edited by D.M. Gallo and R.M. Porter. V, 117 pages. 1983.

Vol. 972: Nonlinear Filtering and Stochastic Control. Proceedings, 1981. Edited by S.K. Mitter and A. Moro. VIII, 297 pages. 1983.

Vol. 973: Matrix Pencils. Proceedings, 1982. Edited by B. Kågström and A. Ruhe. XI, 293 pages. 1983.

Vol. 974: A. Draux, Polynômes Orthogonaux Formels – Applications. VI, 625 pages. 1983.

Vol. 975: Radical Banach Algebras and Automatic Continuity. Proceedings, 1981. Edited by J.M. Bachar, W.G. Bade, P.C. Curtis Jr., H.G. Dales and M.P. Thomas. VIII, 470 pages. 1983.

Vol. 976: X. Fernique, P.W. Millar, D.W. Stroock, M. Weber, Ecole d'Eté de Probabilités de Saint-Flour XI – 1981. Edited by P.L. Hennequin. XI, 465 pages. 1983.

Vol. 977: T. Parthasarathy, On Global Univalence Theorems. VIII, 106 pages. 1983.

Vol. 978: J. Ławrynowicz, J. Krzyż, Quasiconformal Mappings in the Plane. VI, 177 pages. 1983.

Vol. 979: Mathematical Theories of Optimization. Proceedings, 1981. Edited by J.P. Cecconi and T. Zolezzi. V, 268 pages. 1983.

Vol. 980: L. Breen. Fonctions thêta et théorème du cube. XIII, 115 pages. 1983.

Vol. 981: Value Distribution Theory. Proceedings, 1981. Edited by I. Laine and S. Rickman. VIII, 245 pages. 1983.

Vol. 982: Stability Problems for Stochastic Models. Proceedings, 1982. Edited by V.V. Kalashnikov and V.M. Zolotarev. XVII, 295 pages. 1983.

Vol. 983: Nonstandard Analysis-Recent Developments. Edited by A.E. Hurd. V, 213 pages. 1983.

Vol. 984: A. Bove, J.E. Lewis, C. Parenti, Propagation of Singularities for Fuchsian Operators. IV, 161 pages. 1983.

Vol. 985: Asymptotic Analysis II. Edited by F. Verhulst. VI, 497 pages. 1983.

Vol. 986: Séminaire de Probabilités XVII 1981/82. Proceedings. Edited by J. Azéma and M. Yor. V, 512 pages. 1983.

Vol. 987: C.J. Bushnell, A. Fröhlich, Gauss Sums and p-adic Division Algebras. XI, 187 pages. 1983.

Vol. 988: J. Schwermer, Kohomologie arithmetisch definierter Gruppen und Eisensteinreihen. III, 170 pages. 1983.

Vol. 989: A.B. Mingarelli, Volterra-Stieltjes Integral Equations and Generalized Ordinary Differential Expressions. XIV, 318 pages. 19[illegible]

Vol. 990: Probability in Banach Spaces IV. Proceedings, 19[illegible] Edited by A. Beck and K. Jacobs. V, 234 pages. 1983.

Vol. 991: Banach Space Theory and its Applications. Proceedings, 1981. Edited by A. Pietsch, N. Popa and I. Singer. X, 302 pages. 1983.

Vol. 992: Harmonic Analysis, Proceedings, 1982. Edited by G. Mauceri, F. Ricci and G. Weiss. X, 449 pages. 1983.

Vol. 993: R.D. Bourgin, Geometric Aspects of Convex Sets with the Radon-Nikodým Property. XII, 474 pages. 1983.

Vol. 994: J.-L. Journé, Calderón-Zygmund Operators, Pseudo-Differential Operators and the Cauchy Integral of Calderón. VI, [illegible] pages. 1983.

Vol. 995: Banach Spaces, Harmonic Analysis, and Probability Theory. Proceedings, 1980–1981. Edited by R.C. Blei and S.J. Sidney. V, 173 pages. 1983.

Vol. 996: Invariant Theory. Proceedings, 1982. Edited by F. Gherardelli. V, 159 pages. 1983.

Vol. 997: Algebraic Geometry – Open Problems. Edited by C. Ciliberto, F. Ghione and F. Orecchia. VIII, 411 pages. 1983.

Vol. 998: Recent Developments in the Algebraic, Analytical, and Topological Theory of Semigroups. Proceedings, 1981. Edited by K.H. Hofmann, H. Jürgensen and H.J. Weinert. VI, 486 pages. 19[illegible]

Vol. 999: C. Preston, Iterates of Maps on an Interval. VII, 205 pages. 1983.

Vol. 1000: H. Hopf, Differential Geometry in the Large, VII, 184 pages. 1983.

Vol. 1001: D.A. Hejhal, The Selberg Trace Formula for PSL(2,[illegible] Volume 2. VIII, 806 pages. 1983.

Vol. 1002: A. Edrei, E.B. Saff, R.S. Varga, Zeros of Sections of Power Series. VIII, 115 pages. 1983.

Vol. 1003: J. Schmets, Spaces of Vector-Valued Continuous Functions. VI, 117 pages. 1983.

Vol. 1004: Universal Algebra and Lattice Theory. Proceedings, 1[illegible] Edited by R.S. Freese and O.C. Garcia. VI, 308 pages. 1983.

Vol. 1005: Numerical Methods. Proceedings, 1982. Edited by V. Pereyra and A. Reinoza. V, 296 pages. 1983.

Vol. 1006: Abelian Group Theory. Proceedings, 1982/83. Edited by R. Göbel, L. Lady and A. Mader. XVI, 771 pages. 1983.

Vol. 1007: Geometric Dynamics. Proceedings, 1981. Edited by J. [illegible] Jr. IX, 827 pages. 1983.

Vol. 1008: Algebraic Geometry. Proceedings, 1981. Edited by J. [illegible]gachev. V, 138 pages. 1983.

Vol. 1009: T.A. Chapman, Controlled Simple Homotopy Theory [illegible] Applications. III, 94 pages. 1983.

Vol. 1010: J.-E. Dies, Chaînes de Markov sur les permutations. [illegible] 226 pages. 1983.